Con with Hispanic Workers

CONTRACTOR'S EDITION

DEDICATION

This book is dedicated to *mi familia*.

Copyright 2005 Roger Waynick

Published by Cool Springs Press,
a Division of Thomas Nelson, Inc.,
P.O. Box 141000, Nashville, Tennessee 37214

Library of Congress
Cataloging-in-Publication Data is available.
ISBN 1-59186-232-9

First Printing 2005
Printed in the United States of America
10 9 8 7 6 5 4 3 2 1

Managing Editor: Ramona Wilkes
Production Design: Sergio Daniel Daldi and
S. E. Anderson

Cover Design: Unlikely Suburban Design

Cool Springs Press books may be purchased in bulk
for educational, business, fundraising, or sales
promotional use. For information, please email
SpecialMarkets@ThomasNelson.com.

Visit the Thomas Nelson website at
www.ThomasNelson.com and the Cool Springs Press
website at **www.coolspringspress.net**.

Communicating with Hispanic Workers

CONTRACTOR'S EDITION

Trish Rodriguez

COOL SPRINGS PRESS
A Division of Thomas Nelson Publishers
Since 1798

www.coolspringspress.net

ACKNOWLEDGMENTS

Many people played a part in the research and writing of this book, and I am deeply grateful to each of them. First of all, my husband, Tony Rodriguez, was an invaluable link to the construction business and was essential in my research. Bill Lawrence, a partner with Lawrence Bros. LLC, graciously spent hours reviewing and giving feedback on the chapters; and Fernando and Gerardo Arteaga, of Azteca Services, worked diligently as Spanish copyeditors. Additionally, I'd like to thank the following people, each of whom made a valuable contribution to this book:

- Joe Alexander, Landscaping
- Carlos Gomez, Gomez Brothers Masonry
- Harry (Skip) Lawrence, Partner, Lawrence Brothers, LLC
- Steve Lupear, Owner, Majestic Building Group
- Louis Potts, Potts Roofing Company
- Stanley J. Williams, President, Bradley Coatings Inc.
- Tom Atkinson
- Kris Bearrs
- Randall and Amy Goodgame
- Ann and Les Grammer
- Christopher McLaurin
- David McLaurin
- Ed McLaurin
- Carol Pierce Olson
- Jefferson and Juliana Rodriguez
- Leslie McLaurin Saine
- Roger Waynick

Finally, I must add that this book would never have happened without a lot of prayer and the help of angels. I thank God for giving me the ability and resources to do this. To Him be the glory!

TABLE OF CONTENTS

INTRODUCTION

Communicating with Hispanic Workers: Contractor's Edition is an easy-to-use phrase book for anyone in construction who works with or would like to work with Spanish-speaking workers or sub-contractors. If you are not currently using Spanish-speaking workers, ***Communicating with Hispanic Workers: Contractor's Edition*** will help you gain access to the growing Spanish-speaking labor force. If you are already using Spanish-speaking workers, these phrases will help you improve communication and reduce errors caused by the language barrier.

This book was written for residential construction and remodeling, but many of the phrases apply to commercial construction, as well. The Spanish is Latin American, and there is limited use of technical language or complex expressions.

I started working on ***Communicating with Hispanic Workers: Contractor's Edition*** by interviewing my husband, Tony Rodriguez. Tony is from Ecuador and came to the United States as an adult. He now owns a successful remodeling and custom carpentry business. In addition to talking with Tony, I met with several other contractors for their ideas and feedback. Some of the people I interviewed work in new construction, some in remodeling. Some use Spanish-speaking workers, others do not. A few are bilingual, though most speak only English. Each of them offered a unique perspective and provided me with excellent suggestions. For a complete list of those who graciously helped in the research and writing for this book, please see the **Acknowledgments**.

WHY I WROTE THIS BOOK
I am a language teacher, trainer, and consultant, and have worked in the United States and

Ecuador. I have taught Spanish and English as a Second Language to adults and children, written articles on adult teaching and learning, and trained various groups on cultural awareness. In 1999 I married into the construction business. My husband, Tony, has taught me a lot about construction and how hard it can be to find good help... even when you speak the language!

Tony has the distinct advantage of being bilingual and a native Spanish speaker. He has never had to deal with a language barrier in his business. As long as I can remember, though, his English-speaking colleagues have asked for help with Spanish. They often lament, "if I only spoke a little Spanish...," but like so many of us, they don't have time to take a class. Most people just need to ask a few specific questions or communicate something quickly. I wrote this book with that in mind.

I hope you find **Communicating with Hispanic Workers: Contractor's Edition** useful, and that it puts you on the road to better communication and productivity. Who knows, you might even find you really like speaking Spanish and pursue learning even more! Good Luck and ¡*Buena suerte*!

Trish McLaurin Rodriguez

HOW TO USE THIS BOOK

- Each chapter in *Communicating with Hispanic Workers: Contractor's Edition* focuses on an area of construction. Additionally, the **Quick Start Phrases** include the numbers, days of the week, and some common expressions in Spanish.
- If you are hiring laborers, **Employment Phrases** is a good place to start. This chapter includes phrases to discuss payment, working conditions, health, and safety.
- If you are using unskilled helpers or day laborers, the **General Work** chapter includes general phrases associated with many areas of construction.
- The remaining chapters focus on:
  - **Carpentry**
  - **Drywall**
  - **Flooring**
  - **Framing**
  - **Insulation**
  - **Landscaping**
  - **Masonry**
  - **Painting**
  - **Roofing**
  - **Siding**
- Each chapter has a **list of tools and equipment** and a **list of materials**, followed by the **phrases**.

THE PHRASES

- The phrases are organized into three sections— **Setting Up**, **Doing the Work** and **Finishing Up**. Within each section, the phrases **are alphabetized** by a **keyword** (the main verb).
- Most of the phrases are commands and requests: "Bring me the ladder," "Install the drywall here," or "Help me," for example. They are short and simple, and do not include complex, technical language— just the basics you need on the job site.

	keyword		stressed syllable

English	*Spanish*	Pronunciation
Do it...	*Hágalo...*	**AH**-ga-low...
...carefully	*...con cuidado*	...con kwee-**DA**-doe
...like this	*...así*	...ah-**SEE**
Don't do it like this.	*No lo haga así.*	no low **AH**-ga ah-**SEE**

SPANISH PRONUNCIATION

- After each phrase is a **Spanish translation** and **pronunciation guide**. Like the phrases in English, the Spanish translations are short and simple.
- To use the **pronunciation guide**, you do not need to know anything about Spanish. The words are broken down into syllables, and the stressed syllables are **bolded** and in **UPPERCASE**. With few exceptions, English sound and letter combinations are used to approximate Spanish sounds.
- As you become more comfortable using this book, learning just a little bit about **Spanish letter sounds** will help you to rely less on the pronunciation guide. Here are a few tips:

  The Spanish pronunciation of most **consonants** (b, c, d, f, g...) is very similar to English, with a few important exceptions:

  ll is pronounced "yeh", as in *million*
  t and **p** are soft, as in *stall* and *space*
  ñ is "ny" as in *onion*
  r is soft, while **rr** is rolled
  v is like a soft "b", as in *February*
  ch is always in *church*

  The **vowels** (a,e,i,o,u) in Spanish each have basically one, distinct sound:

  a- -**ah** as in *father*
  e- -**eh** as in *less*
  i- -**ee** as in *see*
  o- -**oh** as in *bone*
  u- -**oo** as in *boot*

9

HELPING YOUR LISTENER UNDERSTAND YOU

If you receive a confused look when you start trying to communicate using the phrases in this book, don't worry. There are different reasons that this may happen, most of which can be easily remedied. For example:

- **Your listener may not be familiar with one of the words you are using**. Spanish is spoken in more than twenty countries (not including the US!). As with English, there is often more than one name for a given item. "Grout" for example, may be referred to as *lechareada*, *sellador de juntas*, or *masilla*. In the lists of tools and materials, I have included alternate translations for many of the words. You might just need to try a different word.

- **Your listener may not know the Spanish word for an item**. Many Spanish-speakers working in construction in the United States did not work in the industry in their native country. They may have never needed to know the word for "grout." Similarly, some construction products that are used in the United States may not be as widely used in other countries. In many Latin American countries, for example, drywall is not used extensively. Though there are plenty of translations for the word "drywall," some workers may simply know it by the "Spanglish" word, "dri wol."[1]

- **Your listener may not be used to hearing accented, "Gringo Spanish."**[2] In time, you will get more comfortable speaking Spanish, and your workers will get more used to hearing you. In the meantime, it never hurts to use gestures and visual cues. Hold up or point to the grout tray when asking someone to bring you the grout.

---

[1] Spanglish—spoken language with combines and borrows from both English and Spanish

[2] Gringo—Latin American slang expression for "North American foreigner"

AND FINALLY...

In addition to using Spanish words and phrases, there are other things that will improve your communication skills in Spanish:

1. **Be polite.** Demonstrating politeness and a degree of formality can go a long way when talking with native Spanish speakers. In Spanish, a person can be addressed using either the *tú* or *usted* form of the verb. *Tú* and *usted* both mean "you" in English, but using *usted* is a sign of courtesy and respect. *Tú* is more commonly used between close friends or when talking to a child. I use the *usted* form exclusively in this book.

   Secondly, I cannot over-stress the importance of saying "thank you," "please," "you're welcome," and "excuse me"—or should I say, *gracias* (gra-**SEE**-ahs), *por favor* (pore **FA**-bore), *de nada* (dey **NA**-da), and *perdón* (pear-**DOAN**).

2. **Take It Slow**. Spoken Spanish may seem very fast. To be understood, do not feel like you have to speak Spanish quickly. It is more important to speak clearly and work on putting the emphasis on the correct syllable.

3. **Remember—Not Everyone Is From Mexico.** Though many Spanish-speaking people in construction are from Mexico, there are more and more people here everyday from other Spanish-speaking countries. Like using the *usted* form and starting your requests with *por favor* ("please"), finding out a bit about the Spanish-speaking people you are working with makes everything go smoother.

4. **Make Sure They Got It.** Pay attention to your listener's reaction. You will soon learn to detect that "deer in the headlights" look that suggests he didn't understand. Similarly, a lot of *si, si, si-ing* ("yes, yes, yes-ing") might be just a polite way of encouraging you, rather than a real response. Repeat yourself, speak slowly, don't

forget visual cues and gestures, and watch to make sure your instructions are followed.

5. **Relax and Have Fun.** You will make mistakes as you begin speaking Spanish. Try to laugh it off and learn from it. There's no need to take yourself too seriously. Your listener may be struggling to learn English and will be sympathetic to your attempts!

Quick Start Phrases

English	Spanish	Pronunciation
Excuse me	*Perdón*	pear-**DOAN**
	Disculpe	diss-**COOL**-pey
I don't understand.	*No entiendo.*	No en-**TEE-EN**-doe.
Please	*Por favor*	pore fa-**BORE**
Repeat that.	*Repita.*	rey-**PEE**-ta.
Thank you.	*Gracias.*	gra-**SEE**-ahs.
You're welcome	*De nada*	dey **NA**-da

DAYS OF THE WEEK

English	Spanish	Pronunciation
Sunday	*domingo*	doe-**MEEN**-go
Monday	*lunes*	**LOO**-nes
Tuesday	*martes*	**MAR**-tehs
Wednesday	*miércoles*	mee-**ER**-co-leys
Thursday	*jueves*	**HWEY**-bes

14 *Quick start phrases*

English	Spanish	Pronunciation
Friday	*viernes*	**BEE-AIR**-nes
Saturday	*sábado*	**SA**-ba-doe
Today	*hoy*	**OHY**
Tomorrow	*mañana*	ma-**NYA**-na
Day after tomorrow	*pasado mañana*	pa-**SA**-doe ma-**NYA**-na
Yesterday	*ayer*	ah-**YER**

NUMBERS

English	Spanish	Pronunciation
1. one	*uno*	**OO**-no
2. two	*dos*	dose
3. three	*tres*	trace
4. four	*cuatro*	**KWA**-tro
5. five	*cinco*	**SEEN**-co

English	Spanish	Pronunciation
6. six	*seis*	seys
7. seven	*siete*	**SEE**-eh-tey
8. eight	*ocho*	**OH**-cho
9. nine	*nueve*	**NEW**-eh-bey
10. ten	*diez*	**DEE**-ace
15. fifteen	*quince*	**KEEN**-sey
20. twenty	*veinte*	**BEN**-tey
30. thirty	*treinta*	**TREN**-ta
40. forty	*cuarenta*	kwa-**REN**-ta
50. fifty	*cincuenta*	seen-**KWEN**-ta
100. one hundred	*cien*	**SEE**-en
101. one hundred one	*ciento y uno*	**SEE**-en-toe ee **OO**-no
125. one hundred twenty-five	*ciento veinticinco*	**SEE**-en-toe ben-tee-**SEEN**-co

INSTRUCTIONS

English	Spanish	Pronunciation
Be careful.	*Tenga cuidado.*	**TEN**-ga kwee-**DA**-doe.
Begin here.	*Empiece aquí.*	em-**PEE-EH**-sey ah-**KEY**.
Build	*Construya*	con-**STREW**-ya
Clean.	*Limpie*	**LEEM**-pee-eh
Come here.	*Venga.*	**BEN**-ga.
Do it like this.	*Hágalo así.*	**AH**-ga-low ah-**SEE**.
Don't do it like this.	*No lo haga así.*	**NO** low **AH**-ga ah-**SEE**.
Help him.	*Ayúdele.*	Ah-**YOU**-dey-ley.
Help me.	*Ayúdeme.*	Ah-**YOU**-dey-mey.
Install	*Instale*	een-**STA**-ley
Paint	*Pinte*	**PEEN**-tey
Remove	*Quite*	**KEY**-tey

Quick start phrases 17

English	Spanish	Pronunciation
Show me.	Muéstreme.	MWES-trey-mey.
Take a break.	Descanse.	des-CAN-sey.
Unload it.	Descárguelo.	des-CAR-gey-low.

GREETINGS & INTRODUCTIONS

English	Spanish	Pronunciation
Good afternoon	Buenas tardes	BWHEY-nas TAR-deys
Good evening	Buenas tardes	BWHEY-nas TAR-deys
Good morning	Buenos días	BWHEY-nos DEE-ahs
Good night	Buenas noches	BWHEY-nas NO-cheys
Hello	Hola	OH-la
I'm the boss.	Soy el jefe. (masculine) Soy la jefa. (feminine)	soy el HEY-fey. soy la HEY-fa.
I'm (name).	Me llamo _____.	mey YA-mo _____.

English	Spanish	Pronunciation
Pleased to meet you.	Mucho gusto.	MOO-cho GOOSE-toe.
What is your name?	¿Cómo se llama?	CO-mo sey YA-ma?
Where is your supervisor?	¿Dónde está su supervisor?	DON-dey es-TA sue sue-pear-bee-SORE?

WORK SCHEDULES

English	Spanish	Pronunciation
Can you work late?	¿Puede trabajar tiempo extra?	PWHEY-dey tra-ba-har TEE-EM-poe ECKS-tra?
Can you work tomorrow?	¿Puede trabajar mañana?	PWHEY-dey tra-ba-HAR ma-NYA-na?
Come at 8 o'clock.	Venga a las ocho.	BEN-ga ah las OH-cho.
...7 o'clock.	...las siete.	... las SEE-eh-tey.
No work tomorrow.	No hay trabajo mañana.	no eye tra-BA-hoe ma-NYA-na.

Carpentry

Tools & Equipment

English	Spanish / Pronunciation
Air Gun	*la pistola de aire*
	la pees-**TOLL**-ah dey **EYE**-rey
Air Hose	*la manguera de aire*
	la man-gey-ra dey **EYE**-rey
Chalk Line	*el marcador de líneas*
	el mar-ka-**DOOR** dey **LEE**-ney-ahs
	la tira líneas
	la **TEE**-ra **LEE**-ney-ahs
Chisel	*el formón*
	el fore-**MON**
	el cincel
	el see-**CELL**
Clamp	*la abrazadora*
	la ah-bra-sa-**DOE**-ra
Compressor	*el compresor*
	el com-prey-**SORE**

English	Spanish / Pronunciation
Crow Bar	*la pata de cabra*
	la **PA**-ta dey **CA**-bra
	la barra
	la **BA**-rra
Drill	*el taladro*
	el ta-**LA**-drow
Drill Bit	*la broca*
	la **BRO**-ka
Extension Cord	*la extensión eléctrica*
	la ex-ten-see-**ON** eh-**LECK**-tree-ca
Hammer	*el martillo*
	el mar-**TEE**-yo
Ladder	*la escalera*
	la es-ca-**LEY**-ra
Level	*el nivel*
	el nee-**BELL**

English	Spanish / Pronunciation
Nail Gun	*la pistola de clavos*
	la pees-**TOE**-la de **CLA**-bose
	la clavadora automática
	la cla-ba-**DOE**-rah ow-toe-**MA**-tee-ca
Nail Puller	*el saca clavos*
	el **SA**-ka **CLA**-bos
Router	*la fresadora*
	la fre-sa-**DOE**-rah
	la ranuradora
	la ra-new-ra-**DOE**-rah
Sander	*la lijadora*
	la lee-ha-**DOE**-rah
Saw, Circular	*la sierra circular*
	la **SEE-EH**-rrah seer-**COO**-lar
Saw, Hand	*la sierra de mano*
	la **SEE-EH**-rrah dey **MA**-no
	el serrucho
	el sey-**RREW**-cho

English	Spanish / Pronunciation
Saw, Jig	*la sierra vaivén*
	la **SEE-EH**-rrah bye-ee-**BEN**
Saw, Mitre	*la sierra de retroceso*
	la **SEE-EH**-rrah dey re-tro-**SEY**-so
Saw, Saber	*la caladora eléctrica*
	la ca-la-**DOE**-rah eh-**LECK**-tree-ka
Saw, Table	*la cortadora de mesa*
	la core-ta-**DOE**-rah dey **MEY**-sa
Screwdriver	*el desarmador*
	el des-are-ma-**DOOR**
	el destornillador
	el des-tor-nee-yah-**DOOR**
Screwdriver, Phillips Head	*el desarmador estrella*
	el des-are-ma-**DOOR** es-**TREY**-ya
	el desarmador punta cruz
	el des-are-ma-**DOOR POON**-ta cruise

Materials

English	Spanish / Pronunciation
Silicon Gun	*la pistola para silicón*
	la pees-**TOE**-la **PA**-ra see-lee-**CON**
Square	*la escuadra*
	la es-**KWA**-dra
Tape Measure	*la cinta métrica*
	la **SEEN**-tah **MEY**-tree-ka
	el metro
	a el **MEY**-trow

English	Spanish / Pronunciation
T-Square	*la regla T*
	la **REY**-gla tey
Utility Knife	*la navaja*
	el cuchillo multiuso
	el coo-**CHEE**-yo mool-tee-**OO**-so

English	Spanish / Pronunciation
2x4	*dos por cuatro*
	dose pore **KWA**-tro
2x6	*dos por seis*
	dose pore seys
2x8	*dos por ocho*
	dose pore **OH**-cho

English	Spanish / Pronunciation
Baseboard	*la barrera*
	la ba-**RREY**-ra
Beam	*la viga*
	la **BEE**-ga
Biscuits	*las galletas*
	las ga-**YE**-tas

English	Spanish / Pronunciation
Blades	*la sierra de discos*
	la **SEE-EH**-rrah dey **DEES**-cos
	las cuchillas
	las coo-**CHEE**-yas
Bolts	*los pernos*
	los **PEAR**-nos
Column	*la columna*
	la co-**LOOM**-na
Doorknob	*la cerradura*
	la sey-rra-**DO**-ra
Epoxy	*la masilla plástica*
	la ma-**SEE**-ya **PLAS**-tee-ka
Glue	*el pegamento*
	el pey-ga-**MEN**-toe
Hardware	*la ferretería*
	la fe-rrey-tey-**REE**-ah

English	Spanish / Pronunciation
Helpers	*los ayudantes*
	los ah-you-**DAN**-teys
Hinges	*las bisagras*
	las bee-**SA**-gras
Joist	*la viga*
	la **BEE**-ga
Lockset	*la cerradura*
	la sey-rra-**DO**-ra
Lumber	*la madera*
	la ma-**DEY**-rah
Lumber, Pressure-treated	*la madera procesada*
	la ma-**DEY**-rah pro-sey-**SAH**-da
Molding	*la moldura*
	la mol-**DO**-rah
Molding, Crown	*la moldura de corona*
	la mol-**DO**-rah dey co-**ROW**-na

English	Spanish / Pronunciation
Nails	*los clavos*
	los **CLA**-bos
Nuts	*las tuercas*
	las **TWHERE**-kas
Particle Board	*la madera enchapada*
	la ma-**DEY**-rah en-cha-**PA**-da
Plywood	*la hoja de madera*
	la **OH**-ha dey ma-**DEY**-rah
Sandpaper	*la lija*
	la **LEE**-ha
Screws	*los tornillos*
	los tore-**NEE**-yos
Screws, Galvanized	*los tornillos galvanizados*
	los tore-**NEE**-yos gal-ban-nee-**SA**-dose

English	Spanish / Pronunciation
Shims	*las cuñas*
	las **COO**-nyas
Studs	*los montantes*
	los mon-**TAN**-teys
Tool Box	*la caja de herramientas*
	la **CA**-ha dey eh-rra-**MEE-EN**-tas
Trim	*el tapamarcos*
	el ta-pa-**MAR**-cose
Washer	*la arandela*
	la ah-ran-**DEY**-la
Wood	*la madera*
	la ma-**DEY**-rah
Wood Putty	*la masilla para madera*
	la ma-**SEE**-ya **PA**-ra ma-**DEY**-ra

Common Phrases for Setting Up

English	Spanish	Pronunciation
Begin...	Empiece...	em-PEA-A-sey...
...here.	...aquí.	...ah-KEE.
...in this room.	...en este cuarto.	...en ES-tey QUAR-toe.
Carry this.	Lleve esto.	YEA-bey ES-toe.
Do it like this.	Hágalo así.	AH-ga-low ah-SEE.
Don't do it like this.	No lo haga así.	no low AH-ga ah-SEE.
What else do you need?	¿Qué más necesita?	kay mass ne-sey-SEE-ta?
Show me what you need.	Muéstreme que necesita.	MWES-trey-mey kay ne-sey-SEE-ta.
Unload the truck.	Descargue el camión.	des-CAR-gey el ka-me-OWN.
Don't waste materials.	No gaste materiales.	No GAS-tey ma-tey-ree-ALL-ess.

Common Phrases for Doing the Work

English	Spanish	Pronunciation
Bring me the nails.	*Tráigame clavos.*	TRY-ga-mey **CLA**-bos.
...the hammer.	*...el martillo.*	...el mar-**TEE**-yo.
Cut it straight.	*Corte derecho.*	CORE-tey dey-**REY**-cho.
Cut this at 45 degrees.	*Corte en cuarenta y cinco grados.*	CORE-tey en kwa-**REN**-ta eee **SEEN**-co **GRA**-dose.
Cut with the circular saw.	*Corte con la sierra circular.*	CORE-tey con la **SEE-EH**-rrah seer-coo-**LAR**.
Fill in the holes with caulk.	*Coloque la masilla en los agujeros.*	co-**LOW**-kay la ma-**SEE**-ya en los ah-goo-**HARE**-ohs.
We'll finish this...	*Terminamos este...*	tear-mee-**NA**-mos **ES**-tey...
...today.	*...hoy.*	...ohy.
...tomorrow.	*...mañana.*	...ma-**NYA**-na.
Glue the wood.	*Use pegamento en la madera.*	OO-sey pey-ga-**MEN**-toe en la ma-**DEY**-ra.
Do you need help?	*¿Necesita ayuda?*	ney-sey-**SEE**-ta ah-**YOU**-da?
Get someone to help you.	*Pida a alguien que le ayude.*	**PEE**-da ah al-**GEE**-en kay ley ah-**YOU**-dey.

28 *Carpentry*

English	Spanish	Pronunciation
Help him lift this.	*Ayúdele a levantar esto.*	ah-YOU-dey-ley ah ley-ban-**TAR ES**-toe.
...move this.	*...a mover esto.*	...ah moe-**BEAR ES**-toe.
Help me...	*Ayúdeme...*	ah-YOU-dey-mey...
...lift this.	*...a levantar esto.*	...ah ley-ban-**TAR ES**-toe.
...move this.	*...a mover esto.*	...ah moe-**BEAR ES**-toe.
Hold it there...	*Sostenga allí...*	sos-**TEN**-ga ah-**YEE**.
...while I nail it.	*...mientras yo clavo.*	...**MEE-EN**-tras yo **CLA**-bo.
...and nail it.	*...y clave.*	...ee **CLA**-bey.
Install the cabinets here.	*Instale los gabinetes aquí.*	een-**STA**-ley los ga-**BEE**-ney-teys ah-**KEY**.
Install...	*Instale...*	een-**STA**-ley...
...the door knob.	*...la cerradura.*	...la sey-rra-**DO**-ra.
...the hinges.	*...las bisagras.*	...las bee-**SA**-gras.
Level it.	*Nivélelo.*	ni-**BEY**-ley-low.
Make sure it's level.	*Chequée el nivel.*	che-kay-**EH** el nee-**BELL**.

English	Spanish	Pronunciation
Lower it...	*Bájela...*	BA-hey-la...
...a little.	*...un poco.*	...oon **POE**-co.
...an inch.	*...una pulgada.*	...**OO**-na pull-**GA**-da.
Mark it with a pencil.	*Marque con el lápiz.*	MAR-kay con el **LA**-peace.
Measure twice, cut once.	*Mida dos veces, corte una sola vez.*	MEE-da dose **BEY**-seys, **CORE**-tey **OO**-na **SO**-la bes.
We need...	*Necesitamos...*	ney-sey-see-**TA**-mos...
...more nails.	*...más clavos.*	...mass **CLA**-bos.
...more molding.	*...más moldura.*	...mass mol-**DO**-rah.
Make sure the wall is plumb.	*Instale la pared a plomo.*	een-**STA**-ley la pa-**RED** ah PLO-moe.
Raise it a little.	*Levántela un poco.*	le-**BAN**-tey-la oon **POE**-co.
Sand with 220 sandpaper.	*Lije con lija dos veinte.*	LEE-hey con **LEE**-ha dose **BEEN**-tey.
...150...	*...ciento cincuenta.*	...**SEE-EN**-toe seen-**KWEN**-ta.
...100...	*...cien.*	...**SEE**-en.
... 80...	*...ochenta.*	...oh-**CHIN**-ta.

30 Carpentry

English	Spanish	Pronunciation
Sand with the...	*Lije con...*	LEE-hey con...
...square sander.	*...la lijadora cuadrada.*	...la lee-ha-**DO**-rah kwa-**DRA**-da.
...circular sander.	*...la lijadora circular.*	...la lee-ha-**DO**-rah seer-coo-**LAR**.
Use ear plugs.	*Use los protectores de oídos.*	OO-sey los pro-teck-**TORE**-es dey oh-**EE**-dos.
Use screws here.	*Use tornillos aquí.*	OO-sey tore-**NEE**-yos ah-**KEY**.
...nails...	*...clavos...*	...**CLA**-bose.
...a 2x10...	*...un dos por diez...*	...oon dose pore **DEE**-ace.

Common Phrases for Finishing Up

English	Spanish	Pronunciation
Clean up the dust.	*Limpie el polvo.*	LEEM-pea-eh el **POLE**-bo.
Clean the tools.	*Limpie las herramientas.*	LEEM-pea-eh las eh-rra-**MEE-EN**-tas.
Please do this again.	*Por favor hágalo otra vez.*	pore fa-**BORE** AH-ga-low **OH**-tra bes.
When will you finish?	*¿Cuando terminará?*	**KWAN**-doe tear-mee-nah-**RA**?

English	Spanish	Pronunciation
Fix this.	*Arregle esto.*	ah-**RREG**-ley **ES**-toe.
Pick up the tools.	*Recoja las herramientas.*	rey-**KO**-ha las eh-rra-**MEE-EN**-tas.
...trash.	*...la basura.*	...la ba-**SUE**-ra.
...leftover materials.	*...el resto de los materiales.*	...el **RESS**-toe dey los ma-tey-ree-**ALL**-ess.
Put everything in the truck.	*Ponga todo en el camión.*	**PONE**-ga **TOE**-doe en el ca-mee-**OWN**.
...the materials...	*...los materiales...*	... los ma-tey-ree-**ALL**-ess...
...the tools...	*...las herramientas...*	... las eh-rra-**MEE-EN**-tas...
Put it in the...	*Póngalo en...*	**PONE**-ga-lo en...
...trash.	*...la basura.*	...la ba-**SUE**-ra.
...truck.	*...el camión.*	...el ca-me-**OWN**.
Sweep the floors.	*Barra los pisos.*	**BA**-rra los **PEE**-sos.
Vacuum in here.	*Aspire por aquí.*	as-**PEE**-rey pore ah-**KEY**.
...up the dust.	*...el polvo.*	...el **POLE**-bow.

Drywall

Tools & Equipment

English	Spanish / Pronunciation
Bucket	*la cubeta*
	la coo-**BEY**-ta
Caulk Gun	*la pistola para masilla*
	la pees-**TOE**-la **PA**-ra ma-**SEE**-ya
Corner Tool	*la herramienta de esquina*
	la eh-rra-**MEE-EN**-ta dey es-**KEY**-na
Drill, Electric	*el taladro eléctrico*
	el ta-**LA**-drow eh-**LECK**-tree-ko
	el atornillador
	el ah-tore-nee-ya-**DOOR**
Drop Cloth	*la manta (para proteger al piso)*
	la **MAN**-ta (**PA**-ra pro-**TEY**-hair al **PEE**-so)
Dust Mask	*la máscara*
	la **MAS**-ka-ra
Hammer	*el martillo*
	el mar-**TEE**-yo

English	Spanish / Pronunciation
Jointer Plane	*el cepillo de mano*
	el sey-**PEE**-yo dey **MA**-no
Ladder	*la escalera*
	la es-ca-**LEY**-ra
Level	*el nivel*
	el nee-**BELL**
Lift, Drywall	*el elevador de tablarroca*
	eh-ley-ba-**DOOR** dey ta-bla-**RRO**-ka
Mud Pan	*el recipiente para la mezcla*
	la rey-see-**PEE-YEN**-tey **PA**-ra la **MESS**-cla
	la bandeja
	la ban-**DEY**-ha
Pencil	*el lápiz*
	el **LA**-peace
Sander	*la lijadora*
	la lee-ha-**DOE**-rah

English	Spanish / Pronunciation
Saw, Drywall	*la sierra para planchas de yeso* la **SEE-EH**-rra **PA**-ra **PLAN**-chas dey **YEA**-so
Scaffold	*el andamio* el an-**DA**-mee-oh
Scissors	*las tijeras* las tee-**HEY**-ras
Screw Gun	*la pistola de tornillos* la pees-**TOE**-la dey tore-**NEE**-yos
Stilts	*los zancos* los **SAN**-kos
Tape Measure	*la cinta métrica* la **SEEN**-tah **MEY**-tree-ka *el metro* el **MEY**-trow

English	Spanish / Pronunciation
Tarp	*la lona* la **LOW**-na
Trowel	*la espátula* la es-**PA**-too-la
T-Square	*la regla T* la **REG**-la **TEY**
	la escuadra la es-**QUA**-dra
Utility Knife	*el cuchillo multiuso* el koo-**CHEE**-yo mool-tee-**OO**-so
	la navaja la na-**BA**-ha

Materials

English	Spanish / Pronunciation
Caulk	*la masilla* la ma-**SEE**-ya
Corner Bead, Metal	*el esquinero de metal* el es-key-**NEY**-row dey **MEY**-tal
Corner Bead, Plastic	*el esquinero de plástico* el es-key-**NEY**-row dey **PLAS**-tee-ko
Drywall (sheet of)	*la tablarroca* la ta-bla-**RRO**-ka *la plancha de yeso* la **PLAN**-cha dey **YEA**-so
Joint Compound	*el compuesto para juntas* el com-**PWES**-toe **PA**-ra **HOON**-tas *la masilla* la ma-**SEE**-ya

English	Spanish / Pronunciation
Mud	*la masilla* la ma-**SEE**-ya
Nails	*los clavos* los **CLA**-bos
Sandpaper	*la lija* la **LEE**-ha
Screws	*los tornillos* los tore-**NEE**-yos
Sheetrock	*la tablaroca* la ta-bla-**RRO**-ka
Tape, Fiberglass Mesh	*la malla de fibra de vidrio* la **MA**-ya dey **FEE**-bra de **BEED**-ree-oh
Tape, Paper	*la cinta de papel* la **SEEN**-ta dey pa-**PELL**

Common Phrases for Setting Up

English	Spanish	Pronunciation
Begin...	Empiece...	em-PEA-A-sey...
...here.	...aquí.	...ah-**KEE**.
...in this room.	...en este cuarto.	...en **ES**-tey **QUAR**-toe.
Carry this.	Lleve esto.	YEA-bey **ES**-toe.
Cover the floor with tarp.	Cubra los pisos con la lona.	COO-bra los **PEE**-sos kon la **LOW**-na.
Hang dry wall in this room.	Coloque planchas de yeso en este cuarto.	ko-LOW-kay **PLAN**-chas de **YEA**-so en **ES**-tey **QUAR**-toe.
...in all the rooms.	...en todas los cuartos.	...en **TOE**-das los **QUAR**-toes.
...on this wall.	...en esta pared.	...en **ES**-ta pa-**RED**.
Hang the ceiling sheets first.	Coloque las planchas del techo primero.	ko-LOW-kay las **PLAN**-chas del **TEY**-cho pree-**MEY**-ro.
Hang the drywall with...	Coloque la tablarroca con...	ko-LOW-kay la ta-bla-**RRO**-ka kon...
...screws.	...tornillos.	...tore-**NEE**-yos.
...nails.	...clavos.	...**CLA**-bos.

English	Spanish	Pronunciation
Hang the sheets...	Coloque las planchas...	ko-LOW-kay las PLAN-chas.
...vertically.	...verticales.	...bear-tee-KAL-ess.
...horizontally.	...horizontales.	...oh-ree-sawn-TAL-ess.
Measure twice, cut once.	Mida dos veces, corte una sola vez.	MEE-da dose BEY-seys, CORE-tey OO-na SO-la bes.
We need...	Necesitamos...	ney-sey-see-TA-mos...
...4x8 (foot) sheets.	...planchas de cuatro por ocho (pies).	...PLAN-chas dey QUA-tro pore OH-cho (PEE-ehs).
...4x12...	...cuatro por doce...	...QUA-tro pore DOE-sey...
You need...	Necesita...	ney-sey-SEE-ta...
...the proper tools.	...herramientas adecuadas.	...eh-rra-MEE-EN-tas ah-dey-KWA-das.
...a new blade.	...una cuchilla nueva.	...OO-na coo-CHEE-ya NEW-eh-ba.
We need ten sheets (of drywall).	Necesitamos diez planchas (de yeso)...	ney-sey-see-TA-mos DEE-ace PLAN-chas (dey YEA-so).
...twenty (more)...	...veinte (más)...	...BEN-tey (mass)...
Don't waste materials.	No gaste materiales.	No GAS-tey ma-tey-ree-ALL-ess.

Common Phrases for Doing the Work

English	Spanish	Pronunciation
Be careful not to break the corners.	*Cuidado con romper las esquinas.*	kwe-DA-doe con ROM-pear las es-KEY-nas.
Bring me...	*Tráigame...*	TRY-ga-mey...
...a sheet (of drywall).	*...una plancha (de yeso).*	...OO-na PLAN-cha (dey YEA-so).
...the mud.	*...la masilla.*	...la ma-SEE-ya.
...the paper.	*...el papel.*	...el pa-PELL.
Cut around the...	*Corte alrededor de...*	CORE-tey al-rey-dey-DOOR dey...
...electrical outlets.	*...las cajas eléctricas.*	...las KA-has eh-LECK-tree-kas.
...air conditioning.	*...el aire-acondicionado.*	...el EYE-rey ah-kon-dee-see-on-NA-doe.
Give me...	*Déme...*	DEY-mey...
...another sheet.	*...otra plancha.*	...OH-tra PLAN-cha.
...the trowel.	*...la espátula.*	...la es-PA-too-la.
...the paper.	*...el papel.*	...el pa-PELL.
...more mud.	*...mas masilla.*	...mass ma-SEE-ya.
Hang drywall in the ceiling.	*Coloque las planchas en el techo.*	co-LOW-kay las PLAN-chas en el TEY-cho.

English	Spanish	Pronunciation
Hang this panel.	*Coloque esta plancha.*	co-LOW-kay **ES**-ta **PLAN**-cha.
Hold it...	*Sostenga...*	sos-**TEN**-ga...
...steady.	...*fíjamente.*	...fee-ha-**MEN**-tey.
...here.	...*aquí.*	...ah-**KEY**.
Install the corner beads.	*Instale los esquineros.*	een-**STA**-ley las es-key-**NEY**-ros.
Lift the panel.	*Levante la plancha.*	ley-**BAN**-tey la **PLAN**-cha.
Mud the joints.	*Ponga masilla en las uniones.*	PONE-ga ma-**SEE**-ya en las oo-**NEE-OWN**-ess.
Do you need more...	*¿Necesita más...*	ney-sey-**SEE**-tah mass...
...mud?	...*masilla?*	...ma-**SEE**-ya?
...paper?	...*papel?*	...pa-**PELL**?
I need a sheet of drywall.	*Necesito una plancha de yeso.*	ney-sey-**SEE**-toe **OO**-na **PLAN**-cha dey **YEA**-so.
...two sheets...	...*dos planchas...*	...dose **PLAN**-chas.
...three sheets...	...*tres planchas...*	...treys **PLAN**-chas.
Put the panels here.	*Ponga las planchas aquí.*	PONE-ga las **PLAN**-chas ah-**KEY**.
Put a lid on the mud.	*Tape la masilla.*	TA-pey la ma-**SEE**-ya.

English	Spanish	Pronunciation
Sand so that it is smooth.	*Lije hasta que esté lisa.*	LEE-hey AHS-ta kay es-TEY LEE-sa.
Sand these joints.	*Lije estas planchas.*	LEE-hey ES-tas PLAN-chas.
Don't over-sand.	*No lije demasiado.*	No LEE-hey dey-ma-SEE-AH-doe.
Tape the joints.	*Coloque la cinta en las uniones.*	co-LOW-kay la SEEN-ta ee las oo-NEE-OWN-ess.
Use a round corner bead. ...square...	*Use el esquinero redondo. ...cuadrado.*	OO-sey la es-KEY-ner-oh rey-DON-doe. ...kwa-DRA-doe.
Use the green sheets in the bathroom.	*Use las planchas verdes en el baño.*	OO-sey las PLAN-chas BEAR-deys en el BA-nyo.
Use the square and the utility knife to cut.	*Use la navaja para cortar.*	OO-sey la na-BA-ha PA-ra CORE-tar.
Use three coats of mud.	*Ponga tres capas de masilla.*	PONE-ga treys KA-pas dey ma-SEE-ya.
Use 5/8" sheets here.	*Use las planchas de cinco octavos aquí.*	OO-sey las PLAN-chas dey SEEN-co oc-TA-bos ah-KEY.
Use the special tape for this joint.	*Use la cinta especial para esta unión.*	OO-sey la SEEN-ta es-PEY-see-al para ES-ta oo-nee-ON.
Work more slowly.	*Trabaje más despacio.*	tra-BA-hey mass des-PA-see-oh.

Drywall 41

Common Phrases for Finishing Up

English	Spanish	Pronunciation
Clean up the dust.	*Limpie el polvo.*	LEEM-pea-eh el **POLE**-bo.
Clean the tools.	*Limpie las herramientas.*	LEEM-pea-eh las eh-rra-**MEE-EN**-tas.
Please do this again.	*Por favor hágalo otra vez.*	pore fa-**BORE** AH-ga-low **OH**-tra bes.
When will you finish?	*¿Cuando terminará?*	**KWAN**-doe tear-mee-nah-**RA**?
We'll finish this…	*Terminamos esto…*	tear-mee-**NA**-mos **ES**-toe…
…today.	…*hoy.*	…*ohy.*
…tomorrow.	…*mañana.*	…ma-**NYA**-na.
Fix this.	*Arregle esto.*	ah-**RREG**-ley **ES**-toe.
Pick up the…	*Recoja…*	rey-**KO**-ha…
…tools.	…*las herramientas.*	…las eh-rra-**MEE-EN**-tas.
…trash.	…*la basura.*	…la ba-**SUE**-ra.
…leftover materials.	…*el resto de los materiales.*	…el **RESS**-toe dey los ma-tey-ree-**ALL**-ess.

English	Spanish	Pronunciation
Put the materials in the truck.	Ponga los materiales en el camión.	PONE-ga los ma-tey-ree-ALL-ess en el ca-me-OWN.
...the tools...	...las herramientas...	...las eh-rra-MEE-EN-tas...
Put it in the...	Póngalo en...	PONE-ga-lo en...
...trash.	...la basura.	...la ba-SUE-ra.
...truck.	...el camión.	...el ca-me-OWN.
Sweep the floors.	Barra los pisos.	BA-rra los PEE-sos.
Touch-up the walls.	Chequée las fallas.	che-key-EH las FA-yas.
Vacuum up the dust.	Aspire el polvo.	as-PEE-rey el POLE-bo.

Employment Phrases

Interview Questions

GREETINGS

English	Spanish	Pronunciation
I'm (name).	Me llamo (___).	mey YA-moe (___).
My name is…	Mi nombre es…	mee NOM-brey es…
What is your name?	¿Cómo se llama?	KO-moe sey YA-ma?
Write your name.	Escriba su nombre.	es-CREE-ba sue NOM-brey.
Pleased to meet you.	Mucho gusto.	MOO-cho GOOSE-toe.
Do you speak English?	¿Habla inglés?	AB-la eng-LES?
I speak a little Spanish.	Hablo poco español.	AB-low PO-ko es-PAN-nyol.

ADDRESS

English	Spanish	Pronunciation
What is your address?	¿Cuál es su dirección?	KEW-ALL es sue dee-reck-see-ON?
Write your address.	Escriba su dirección.	es-CREE-ba sue dee-reck-see-ON.

46 Employment phrases

TELEPHONE NUMBER

English	Spanish	Pronunciation
What is your telephone number?	¿Cuál es su número de teléfono?	**KEW-ALL** es sue NEW-mey-row dey tey-**LEY**-fo-no?
Write your telephone number.	Escriba su número de teléfono.	es-**CREE**-ba sue NEW-mey-row dey tey-**LEY**-fo-no.

TOOLS

English	Spanish	Pronunciation
Do you have your own tools?	¿Tiene sus propias herramientas?	**TEE-EN**-ey sues pro-**PEE**-ahs eh-rra-**MEE-EN**-tas?
You need your own tools.	Necesita sus propias herramientas.	ney-sey-**SEE**-ta sues pro-**PEE**-ahs eh-rra-**MEE-EN**-tas.
Without tools, there's no work.	Sin herramientas, no hay trabajo.	seen eh-rra-**MEE-EN**-tas, no eye tra-**BA**-hoe.

TRANSPORTATION

English	Spanish	Pronunciation
Do you have transportation?	¿Tiene transporte?	**TEE-EN**-ey trans-**PORE**-tey?
Can you drive a car?	¿Sabe conducir?	**SA**-bey con-due-**SEER**?
Do you have a driver's license?	¿Tiene licencia de conducir?	**TEE-EN**-ey lee-**SEN**-see-ah dey con-due-**SEER**?

UNION

English	Spanish	Pronunciation
Do you have a union card?	*¿Tiene credencial de la unión?*	**TEEN-EN**-ey crey-den-see-**AL** dey la oon-ee-**ON**?
May I see it please?	*¿Puedo verlo por favor?*	**PWHEY**-doe **BEAR**-low pore fa-**BORE**?
Did the union send you?	*¿Lo mandó la unión?*	low man-**DOE** la oon-ee-**ON**?
May I see the referral?	*¿Puedo ver la hoja de referencia?*	**PWHEY**-doe **BEAR** la **OH**-ha dey rey-fey-**REN**-see-ah?

SUB-CONTRACTORS
(Licenses & Insurance)

English	Spanish	Pronunciation
Do you have a business license?	*¿Tiene licencia de negocio?*	**TEEN-EN**-ey lee-**SIN**-see-ah dey ney-**GO**-see-oh?
...liability insurance?	*...seguro contra daños?*	...sey-**GOO**-ro **CON**-tra **DA**-nyos?
...workers comp. insurance?	*...seguro de indemnización de trabajadores?*	...sey-**GOO**-ro dey en-dem-nee-sa-see-**ON** dey tra-ba-ha-**DOE**-reys?

Paperwork

THE APPLICATION

English	Spanish	Pronunciation
Fill out this application.	*Complete esta solicitud.*	come-**PLEY**-tey **ES**-ta so-lee-see-**TUDE**.
Use a pen.	*Use pluma.*	**OO**-sey PLEW-ma.

TAX FORMS (W-2 and W-9)

English	Spanish	Pronunciation
Please fill out this...	*Complete este...*	come-**PLEY**-tey **ES**-tey
...federal tax form.	*...formulario de impuestos federales.*	...for-moo-**LA**-ree-oh dey eem-**PWES**-toes fey-de-**RA**-leys.
...state tax form.	*...formulario de impuestos estatales.*	...for-moo-**LA**-ree-oh dey eem-**PWES**-toes-es-ta-**TA**-les.

I-9 FORM

English	Spanish	Pronunciation
Please fill out this I-9 form.	*Complete este formulario I-9.*	come-**PLEY**-tey **ES**-tey for-moo-**LA**-ree-oh ee **NEW**-eh-bey.

English	Spanish	Pronunciation
I need to see the identification.	*Necesito ver su identificación.*	ney-sey-**SEE**-toe bear sue ee-dent-tee-fee-ka-see-**ON**.
I need to see the original document.	*Necesito ver el documento original.*	ney-sey-**SEE**-toe bear el doe-coo-**MEN**-toe oh-ree-hee-**NAL**.
They won't accept a copy.	*No se aceptan copias.*	no say ah-**SEP**-tan co-**PEE**-ahs.
Please use one from A or...	*Use una de la A o*	OO-sey **OO**-na dey la **AH** oh
...one each from B and C.	*...una de la B y una de la C.*	...**OO**-na dey **OO**-na dey la **BEY** ee **OO**-na dey la **SEY**.
Without appropriate identification, I can't hire you.	*Sin identificación adecuada, no le puedo emplear.*	seen ee-den-tee-fee-ca-see-**ON** ah-dey-**KWA**-da, no ley **PWHEY**-doe em-pley-**ARE**.

Payment
WAGE

English	Spanish	Pronunciation
You will be paid by the hour.	*Se le va a pagar por hora.*	sey ley ba ah pa-**GAR** pore **OH**-ra.
I pay $____ an hour.	*Pago ____ dólares por hora.*	PA-go ____ **DOE**-lay-ess pore **OH**-ra.

English	Spanish	Pronunciation
Your pay is less taxes and benefits.	*El pago es menos impuestos y beneficios.*	el **PA**-go es **MEY**-nos eem-**PWES**-toes y ben-eh-**FEE**-see-ohs.

PAYDAY

English	Spanish	Pronunciation
Can I pay you with a check?	*¿Le puedo pagar con cheque?*	ley **PWHEY**-doe pa-**GAR** con **CHE**-kay?
You will get paid every two weeks.	*Se le va a pagar cada dos semanas.*	sey ley ba ah pa-**GAR CA**-da dose sey-**MA**-nas.
...at the end of the week.	*...a fin de la semana.*	...ah feen dey la sey-**MA**-na.
...at the end of the month.	*...a fin del mes.*	...ah feen dell mes.
Payday is (every) Friday.	*El día del pago es (cada) viernes.*	el **DEE**-ah dell **PA**-go es (**CA**-da) **BEE**-**AIR**-ness.
...Saturday.	*...sábado.*	...**SA**-ba-doe.
...Sunday.	*...domingo.*	...doe-**MEAN**-go.
...Monday.	*...lunes.*	...**LOO**-ness.
...Tuesday.	*...martes.*	...**MAR**-tess.
...Wednesday.	*...miércoles.*	...mee-**AIR**-co-less.
...Thursday.	*...jueves.*	...**HWEY**-bess.

Hours

English	Spanish	Pronunciation
Take a 15 minute break.	*Tome descanso de quince minutos.*	**TOE**-mey des-**CAN**-so dey **KEEN**-sey mee-**NEW**-toes.
Lunch is 30 minutes.	*El almuerzo es de treinta minutos.*	el al-**MWHERE**-so es dey **TREN**-tah mee-**NEW**-toes.
Be here at 6:30 tomorrow.	*Venga mañana a las seis.*	**BEN**-ga ma-**NYA**-na a a las **SEYS**
...7:00...	*...siete...*	...**SEE-EH**-tey...
...8:00...	*...ocho...*	...**OH**-cho...
Don't be late.	*No llegue tarde.*	no **YEA**-gey **TAR**-dey.
We start work at 7:00.	*Empezamos a las siete.*	em-pey-**SA**-mos ah las **SEE-EH**-tey.
...6:00.	*...seis.*	...**SEYS**.
...6:30.	*...seis y media.*	...**SEYS** ee **MEY**-dee-ah.
...7:30.	*...siete y media.*	...**SEE-EH**-tey ee **MEY**-dee-ah.
We finish at 5:00.	*Terminamos a las cinco.*	ter-mee-**NA**-mos a las **SEE**-co
Call me if you cannot work.	*Si no puede trabajar, llámeme.*	see no **PWHEY**-dey tra-ba-**HAR**, **YA**-mey-mey.
...the supervisor...	*...llame al supervisor...*	...**YA**-mey al sue-pear-bee-**SORE**.
...the office...	*...llame a la oficina...*	...**YA**-mey a la oh-fee-**SEE**-na...

52 *Employment phrases*

English	Spanish	Pronunciation
There's no work tomorrow.	*No hay trabajo mañana.*	no eye tra-BA-ho ma-NYA-na.
Can you work tomorrow?	*¿Puede trabajar mañana?*	PWEY-dey tra-ba-HAR ma-NYA-na?
Can you work late?	*¿Puede trabajar tiempo extra?*	PWEY-dey tra-ba-HAR TEE-EM-poe ECKS-tra?

Safety

HEALTH & INJURIES

English	Spanish	Pronunciation
Call 911.	*Llame al nueve once.*	YA-mey al NEW-eh-bey ON-cey.
Go to the doctor now!	*Vaya al doctor ahora.*	BA-ya all dock-TORE ah-OH-ra.
...clinic.	*...a la clínica.*	...a la CLEE-nee-ka.
Bring me the doctor's report.	*Tráigame el reporte del doctor.*	TRY-ga-mey el rey-PORE-tey dell dock-TORE.
Are you injured?	*¿Se lesionó?* (Mex.)	sey ley-see-on-NO?
	¿Se lastimó?	sey lass-tee-MOE?
Are you sick?	*¿Está enfermo?*	ES-ta en-FAIR-moe?

PROTECTION

English	Spanish	Pronunciation
Hard hat area.	*Use casco en esta área.*	**OO**-sey **KAS**-co en **ES**-ta **AH**-rey-ah.
Here is your safety booklet.	*Aquí está su folleto de seguridad.*	ah-**KEY** **ES**-ta sue fo-**YEA**-toe de sey-goo-ree-**DAD**.
Wear these glasses for protection.	*Use estos lentes para protección.*	OO-sey **ES**-toes **LEN**-tes **PA**-ra pro-teck-see-**ON**.
...this hard hat...	*...este casco...*	...**ES**-tey **CAS**-co...
...these gloves...	*...estos guantes...*	...**ES**-toes **GWAN**-teys...
...ear plugs...	*...tapones de oídos...*	...ta-**PONE**-es dey oh-**EE**-does...

WARNINGS

English	Spanish	Pronunciation
Be careful.	*Tenga cuidado.*	**TEN**-ga kwee-**DA**-doe.
Caution	*Aviso*	ah-**BEE**-so
Danger	*Peligro*	pey-**LEE**-grow
Do not enter.	*No entrar.*	No en-**TRAR**.
Don't move.	*No se mueva.*	no sey **MOO-EH**-ba.

54 *Employment phrases*

English	Spanish	Pronunciation
Duck!	¡Agáchese!	ah-**GA**-che-sey!
Get away!	¡Quítese!	**KEY**-tey-sey!
Get down!	¡Bájese!	**BA**-hey-sey!
Stop!	¡Alto!	**AL**-toe!
	¡Pare!	**PA**-rey!
Watch out!	¡Cuidado!	kwee-**DA**-doe!
	¡Ojos!	**OH**-hoes!
	¡Aguas! (Mex.)	**AH**-gwas!

Work Rules

GENERAL

English	Spanish	Pronunciation
Clean up your lunch trash.	Limpie la basura de la comida.	**LEEM**-pee-eh la ba-**SUE**-ra dey la co-**MEE**-da.
Don't cook in the building.	No cocine dentro del edificio.	No co-**SEE**-ney **DEN**-tro dell eh-dee-**FEE**-see-oh.

English	Spanish	Pronunciation
Don't work in the dark.	*No trabaje en la oscuridad.*	No tra-**BA**-hey en la os-coo-ree-**DAD** .
No drinking.	*No tomar alcohol.*	no toe-**MAR** al-co-**ALL**.
No drug use.	*No usar drogas.*	no oo-**SAR DRO**-gas.
No music.	*No música.*	no **MOO**-see-ka.
No smoking.	*No fumar.*	no foo-**MAR**.
Take your shoes off before you enter.	*Quítese los zapatos antes de entrar.*	**KEY**-tey-sey los sa-**PA**-toes **AN**-teys dey en-**TRAR**.
Turn the volume down.	*Baje el volumen.*	**BA**-hey el bow-loo-**MEN**.

PARKING

English	Spanish	Pronunciation
Park on this side of the street.	*Estacionar a este lado de la calle.*	es-ta-see-oh-**NAR** ah **ES**-tey **LA**-doe dey la **CA**-yea.
Don't park on this side of the street.	*No estacionar a este lado de la calle.*	no es-ta-see-oh-**NAR** a **ES**-tey **LA**-doe dey la **CA**-yea.
Don't park here.	*No estacionar aquí.*	no es-ta-see-oh-**NAR** ah-**KEY**.

RESTROOMS

English	Spanish	Pronunciation
You may use this bathroom.	Puede usar este baño.	PWHEY-de oo-SAR ES-tey BA-nyo.

SUPERVISION

English	Spanish	Pronunciation
I'm the boss.	Soy el jefe. *(masculine)* Soy la jefa. *(feminine)*	soy el HEY-fey. soy la HEY-fa.
I'm the supervisor.	Soy el supervisor. *(masculine)* Soy la supervisora. *(feminine)*	soy el sue-pear-bee-SORE. soy la sue-pear-bee-SO-ra.
(name) is the boss.	_____ es el jefe. *(masculine)* _____ es la jefa. *(feminine)*	_____ es el HEY-fey. _____ es la HEY-fa.
Don't use any tools without asking.	No use ninguna herramienta sin permiso.	no OO-sey neen-GOO-na eh-rra-MEE-EN-ta seen pear-MEE-so.
Use your own tools.	Use sus propias herramientas.	OO-sey sues PRO-pee-ahs eh-rra-MEE-EN-tas.

Flooring

Tools & Equipment

English	Spanish / Pronunciation
Air Hose	*la manguera de aire* la man-**GEY**-ra dey **EYE**-rey
Bucket	*la cubeta* la coo-**BEY**-ta
Chalk Line	*el marcador de líneas* el mar-**KA**-door dey **LEE**-ney-ahs *la tira líneas* la **TEE**-ra **LEE**-ney-ahs
Clamp	*la abrazadora* la ah-bra-sa-**DOE**-ra
Compressor	*el compresor* el com-prey-**SORE**
Drill	*el taladro* el ta-**LA**-drow

English	Spanish / Pronunciation
Extension Cord	*la extensión eléctrica* la ex-ten-see-**ON** eh-**LECK**-tree-ca
Gloves, Rubber	*los guantes de caucho* los **GWAN**-teys dey **COW**-cho
Gloves, Work	*los guantes de trabajo* los **GWAN**-teys dey tra-**BA**-hoe
Grout Float	*la llana para lecharear* la **YA**-na **PA**-ra ley-cha-rey-**ARE**
Hammer	*el martillo* el mar-**TEE**-yo
Knee Pads	*las rodilleras* las row-dee-**YEA**-ras
Level	*el nivel* el nee-**BELL**

English	Spanish / Pronunciation
Mallet	*el mazo* / el **MA**-so
Mixing Paddle	*la paleta* / la pa-**LEY**-ta
Nail Gun	*la pistola de clavos* / la pees-**TOE**-la de **CLA**-bose
	la clavadora automática / la cla-ba-**DOE**-rah ow-toe-**MA**-tee-ca
Pull Bar	*la barra* / la **BA**-rra
Putty Knife	*la espátula* / la es-**PA**-too-la
Sander	*la lijadora* / la lee-ha-**DOE**-rah

English	Spanish / Pronunciation
Saw, Hand	*la sierra de mano* / la **SEE-EH**-rra dey **MA**-no
	el serrucho / el sey-**RREW**-cho
Saw, Jig	*la sierra vaivén* / la **SEE-EH**-rra bye-ee-**BEN**
Saw, Mitre	*la sierra de retroceso* / la **SEE-EH**-rra dey re-tro-**SEY**-so
Saw, Table	*la cortadora de mesa* / la core-ta-**DOE**-rah dey **MEY**-sa
Saw, Tile	*la cortadora de azulejos* / la core-ta-**DOE**-ra dey ah-sue-**LEY**-hoes
Saw, Wet	*la sierra para cortar en húmedo* / la **SEE-EH**-rra **PA**-ra **CORE**-tar en **OO**-mey-doh

English	Spanish / Pronunciation
Tile Cutters	*la tijera para azulejo*
	las tee-**HEY**-ras pay-ra ah-sue-**LEY**-hoe
	la tenaza
	la tey-**NA**-sa
Tile Spacers	*los espaciadores*
	los es-pa-see-ah-**DOE**-reys
Trowel	*la llana*
	la **YA**-na
Trowel, Notched	*la espátula dentada*
	la es-**PA**-too-la den-**TA**-da
Utility Knife	*la navaja*
	la na-**BA**-ha
	el cuchillo multiuso
	el coo-**CHEE**-yo mool-tee-**OO**-so

English	Spanish / Pronunciation
Scissors	*las tijeras*
	las tee-**HEY**-ras
Seam Roller	*el rodillo de juntas*
	el ro-**DEE**-yo dey **WHOON**-tas
Sponge	*la esponja*
	la es-**PONE**-ha
Square	*la escuadra*
	la es-**KWA**-dra
Straight Edge	*la regla*
	la **REY**-gla
Tape Measure	*la cinta métrica*
	la **SEEN**-tah **MEY**-tree-ka
	el metro
	el **MEY**-trow

Materials

English	Spanish / Pronunciation
Adhesive	*el adhesivo* el ad-eh-**SEE**-bow
Backer Board	*la plancha base* la **PLAN**-cha **BA**-sey
Ceramic Tile	*el azulejo* el ah-sue-**LEY**-hoe *la loseta* la low-**SEY**-ta
Cork	*el corcho* el **CORE**-cho
Felt Paper	*el papel fieltro* el pa-**PELL** fee-**EL**-trow
Floor Leveling	*la masilla para nivelar* la ma-**SEE**-ya **PA**-ra nee-bey-**LAR**

English	Spanish / Pronunciation
Foam	*la esponja* la es-**PONE**-ha
Grout	*el sellador de juntas* el sey-ya-**DOOR** dey **WHOON**-tas *la lechareada* la ley-cha-rey-**AH**-da
Hardwood Planks	*las tiras de madera* las **TEE**-ras dey ma-**DEY**-ra
Mineral Spirits	*el diluyente* el dee-lew-**YEN**-tey
Nails	*los clavos* las **CLA**-bows
Plywood	*la hoja de madera* la **OH**-ha dey ma-**DEY**-ra

English	Spanish / Pronunciation
Sand Paper	*la lija*
	la **LEE**-ha
Sealant	*el sellador*
	el se-ya-**DOOR**
Underlayment	*la plancha base*
	la **PLAN**-cha **BA**-sey

English	Spanish / Pronunciation
Vinyl	*el vinilo*
	el bee-**NEE**-low
Wax	*la cera*
	la **SEY**-ra
Weather Barrier	*la barrera de agua y de viento*
	la ba-**RREY**-ra dey **AH**-gwa ee dey **BEE-EN**-toe

Common Phrases for Setting Up

English	Spanish	Pronunciation
Begin...	*Empiece...*	em-PEA-A-sey...
...here.	...*aquí.*	...ah-**KEE**.
...over there.	...*allí.*	...ah-**YEE**.
...in this room.	...*en este cuarto.*	...en **ES**-tey **QUAR**-toe.
Carry this.	*Lleve esto.*	**YEA**-bey **ES**-toe.

English	Spanish	Pronunciation
Clean the floor thoroughly.	*Limpie el piso bien.*	LEEM-pee-eh el **PEE**-so **BEE**-en.
Measure twice, cut once.	*Mida dos veces, corte una sola vez.*	MEE-da dose **BEY**-seys, **CORE**-tey **OO**-na **SO**-la bes
Mix tiles from different boxes.	*Mezcle losetas de varias cajas.*	MESS-cley low-**SEY**-tas dey ba-**REE**-ahs **CA**-has.
...pieces	...*piezas*	...**PEE-EH**-sas.
Put down backboard.	*Ponga la plancha base.*	PONE-ga la **PLAN**-cha **BA**-sey.
...felt paper.	...*el papel fieltro.*	...el pa-**PELL** fee-**EL**- tro.
...plywood.	...*la hoja de madera.*	...la **OH**-ha dey ma-**DEY**-ra.
Put the boxes in the room.	*Ponga las cajas en el cuarto.*	PONE-ga las KA-has en el **KWAR**-toe.
Remove the carpet.	*Quite la alfombra.*	KEY-tey la al-**FOM**-bra.
...the damaged areas.	...*las áreas dañadas.*	...las **AH**-rey-ahs da-**NYA**-das.
...the baseboards.	...*las barrederas.*	...las ba-rrey-**DEY**-ras.
Replace the sub-floor.	*Reemplace la plancha base.*	rey-em-**PLA**-sey la **PLAN**-cha **BA**-sey.
Sweep the floor first.	*Barra el piso primero.*	**BA**-rra el **PEE**-so pree-**MEY**-row.
Unload the truck.	*Descargue el camión.*	des-**CAR**-gey el ka-me-**OWN**.
Don't waste materials.	*No gaste materiales.*	No **GAS**-tey ma-tey-ree-**ALL**-ess.

Common Phrases for Doing the Work

English	Spanish	Pronunciation
Apply a thin coat of mortar.	*Aplique una capa delgada de mortero.*	ah-PLEE-kay **OO**-na **CA**-pa dell-**GA**-da dey more-**TEY**-row.
Be careful with the saw.	*Cuidado con la cortadora.*	kwe-**DA**-doe con la core-ta-**DOE**-ra.
Bring me the hammer.	*Tráigame el martillo.*	TRY-ga-mey el mar-**TEE**-yo.
...the nails.	...*clavos.*	...**CLA**-bos.
...the tile.	...*la loseta.*	...la low-**SEY**-ta.
Cut it straight.	*Corte derecho.*	CORE-tey dey-**REY**-cho.
Cut this at 45 degrees.	*Corte en cuarenta y cinco grados.*	CORE-tey en kwa-**REN**-ta ee **SEEN**-co **GRA**-dose.
Cut with the circular saw.	*Corte con la sierra circular.*	CORE-tey con la **SEE-EH**-rrah seer-coo-**LAR**.
We'll finish this today.	*Terminamos este hoy.*	tear-mee-**NA**-mos **ES**-tey ohy.
...tomorrow.	...*mañana.*	...ma-**NYA**-na.
Follow the wood grain.	*Siga las líneas de la madera.*	SEE-ga las **LEE**-ney-ahs de la ma-**DEY**-ra.

English	Spanish	Pronunciation
Glue the wood.	Use pegamento en la madera.	OO-sey pey-ga-MEN-toe en la ma-DEY-ra.
Do you need help?	¿Necesita ayuda?	ney-sey-SEE-ta ah-YOU-da?
Get someone to help you.	Pida a alguien que le ayude.	PEE-da ah AL-gee-en kay ley ah-YOU-dey.
Help me...	Ayúdeme...	ah-YOU-dey-mey...
...lift this.	...a levantar esto.	...ah ley-ban-TAR ES-toe.
...move this.	...a mover esto.	...ah moe-BEAR ES-toe.
Hold it there...	Sostenga allí...	sos-TEN-ga ah-YEE...
...while I nail it.	...mientras yo clavo.	...MEE-EN-tras yo CLA-bo.
...and nail it.	...y clave.	...ee CLA-bey.
Install the planks like this.	Instale las tiras de madera así.	een-STA-ley las TEE-ras dey ma-DEY-ra ah-SEE.
...the tiles...	...las losetas...	...las low-SEY-tas.
...the vinyl...	...el vinilo...	...el be-NEE-low.
Mark the floor with a pencil.	Marque el piso con el lápiz.	MAR-kay el PEE-so con el LA-peace.
...a chalk line.	...la tira líneas.	...la TEE-ra LEE-ney-ahs.

English	Spanish	Pronunciation
Measure twice, cut once.	Mida dos veces, corte una sola vez.	MEE-da dose **BEY**-seys, **CORE**-tey **OO**-na **SO**-la bes.
We need...	Necesitamos...	ney-sey-see-**TA**-mos...
...more nails.	...más clavos.	...mass **CLA**-bos.
...more molding.	...más moldura.	...mass mol-**DO**-ra.
Sand with...sandpaper.	Lije con liga...	LEE-hey con **LEE**-ha...
...220...	...dos veinte.	...dose **BEEN**-tey.
...150...	...ciento cincuenta.	...**SEE-EN**-toe seen-**KWEN**-ta.
...100...	...cien.	...**SEE**-en.
...80...	...ochenta.	...oh-**CHIN**-ta.
Seal the grout.	Aplique sellador en la lechareada.	ah-**PLEE**-kay sey-ya-**DOOR** en la ley-cha-rey-**AH**-da.
Let the mortar set up overnight.	Deje que el mortero se seque toda la noche.	**DEY**-hey kay el more-**TEY**-row sey **SEY**-kay **TOE**-da la **NO**-che.
Use spacers between tiles.	Use espaciadores entre las losetas.	**OO**-sey es-pa-see-ah-**DOE**-reys **EN**-trey las low-**SEY**-tas.
Use the notched trowel.	Use la espátula dentada.	**OO**-sey la es-**PA**-too-la den-**TA**-da.

Common Phrases for Finishing Up

English	Spanish	Pronunciation
Clean up...	Limpie...	LEEM-pea-eh...
...the dust.	...el polvo.	...el POLE-bo.
...the floor.	...el piso.	...el PEE-so.
Clean the tools.	Limpie las herramientas.	LEEM-pea-eh las eh-rra-MEE-EN-tas.
Please do this again.	Por favor hágalo otra vez.	pore fa-BORE AH-ga-low OH-tra bes.
When will you finish?	¿Cuando terminará?	KWAN-doe tear-mee-nah-RA?
Fix this.	Arregle esto.	ah-RREG-ley ES-toe.
Pick up the...	Recoja...	rey-KO-ha...
...tools.	...las herramientas.	...las eh-rra-MEE-EN-tas.
...trash.	...la basura.	...la ba-SUE-ra.
...leftover materials.	...el resto de los materiales.	...el RESS-toe dey los ma-tey-ree-ALL-ess.
Put the materials in the truck.	Ponga los materiales en el camión.	PONE-ga los ma-tey-ree-ALL-ess en el ca-me-OWN.
...the tools	...las herramientas	...las eh-rra-MEE-EN-tas.

English	Spanish	Pronunciation
Put it in the...	*Póngalo en...*	PONE-ga-lo en...
...trash.	*...la basura.*	...la ba-**SUE**-ra.
...truck.	*...el camión.*	...el ca-me-**OWN**.
...dump truck.	*...el camión de volteo.*	...el ca-me-**OWN** dey bowl-**TEY**-oh.
Sweep the floors.	*Barra los pisos.*	BA-rra los **PEE**-sos.
Vacuum...	*Aspire...*	as-PEE-rey...
...in here.	*...por aquí.*	...pore ah-**KEY**.
...up the dust.	*...el polvo.*	...el **POLE**-bow.

Framing

Tools & Equipment

English	Spanish / Pronunciation
Air Gun	*la pistola de aire*
	la pees-**TOLL**-ah dey **EYE**-rey
Air Hose	*la manguera de aire*
	la man-**GEY**-ra dey **EYE**-rey
Chalk Line	*el marcador de líneas*
	el mar-ka-**DOOR** dey **LEE**-ney-ahs
Chalk Line	*la línea de marcar*
	la **LEE**-nay-ah dey **MAR**-car
Extension Cord	*la extensión eléctrica*
	la ex-ten-see-**ON** eh-**LECK**-tree-ca
Hammer	*el martillo*
	el mar-**TEE**-yo
Ladder	*la escalera*
	la es-ca-**LEY**-ra
Level	*el nivel*
	el nee-**BELL**

English	Spanish / Pronunciation
Nail Gun	*la pistola de clavos*
	la pees-**TOE**-la de **CLA**-bose
	la clavadora automática
	la cla-ba-**DOE**-rah ow-toe-**MA**-tee-ka
Saw, Circular	*la sierra circular*
	la **SEE-EH**-rrah seer-coo-**LAR**
Tape Measure	*la cinta métrica*
	la **SEEN**-tah **MEY**-tree-ka
	el metro
	a el **MEY**-trow
Saw, Radial Arm	*el serrucho radial*
	el sey-**RROO**-cho ra-dee-**ALL**
Saw Horse	*el caballete*
	el ca-ba-**YE**-tey

Materials

English	Spanish / Pronunciation
2x4	*dos por cuatro*
	dose pore **KWA**-tro
2x6	*dos por seis*
	dose pore seys
2x8	*dos por ocho*
	dose pore **OH**-cho
2x10	*dos por diez*
	dose pore **DEE**-ace
Beam	*la viga*
	la **BEE**-ga
Beam, Double	*la doble viga*
	la **DOE**-bley **BEE**-ga
Bolts	*los pernos*
	los **PEAR**-nos

English	Spanish / Pronunciation
Bolts, Anchor	*los pernos de anclaje*
	los **PEAR**-nos dey an-**CLA**-hey
Header	*la doble viga*
	la **DOE**-bley **BEE**-ga
	el cabezal
	el ca-**BEY**-sal
Joist	*la viga*
	la **BEE**-ga
Lumber,	
Pressure-treated	*la madera procesada*
	la ma-**DEY**-rah pro-sey-**SAH**-da
Nails	*los clavos*
	los **CLA**-bos
Nuts	*las tuercas*
	las **TWHERE**-kas

English	Spanish / Pronunciation
Particle Board	*la madera enchapada*
	la ma-**DEY**-rah en-cha-**PA**-da
Plates	*las soleras*
	las so-**LEY**-ras
Plate, Bottom	*la solera inferior*
	la so-**LEY**-ra een-fey-ree-**OR**
Plate, Top	*la solera superior*
	la so-**LEY**-ra sue-pey-ree-**OR**
Plywood	*la hoja de madera*
	la **OH**-ha dey ma-**DEY**-rah

English	Spanish / Pronunciation
Shims	*las cuñas*
	las **COO**-nyas
Studs	*los montantes*
	los mon-**TAN**-teys
Washer	*la arandela*
	la ah-ran-**DEY**-la
Wood	*la madera*
	la ma-**DEY**-rah

Common Phrases for Setting Up

English	Spanish	Pronunciation
Begin...	Empiece...	em-**PEA-A**-sey...
...here.	...aquí.	...ah-**KEE**.
...over there.	...allí.	...ah-**YEE**.
Look at the plans first.	Vea los planos primero.	**BEY**-ah los **PLA**-nos pre-**MEY**-roe.
Measure and mark the floors.	Mida y marque los pisos.	**MEE**-da y **MAR**-kay los **PEE**-sos.
Measure the layout of each wall.	Mida el trazo de cada pared.	**MEE**-day el **TRA**-so dey **CA**-da pa-**RED**.
You need...	Necesita...	ney-sey-**SEE**-ta...
...new tools.	...nuevas herramientas.	...**NEW**-eh-bas eh-rra-**MEE-EN**-tas.
...a new blade.	...una navaja nueva.	...**OO**-na na-**BA**-ha **NEW**-eh-ba.
Put the wood...	Ponga la madera...	**PONE**-ga la ma-**DEY**-ra...
...here.	...aquí.	...ah-**KEY**.
...over there.	...allí.	...ah-**YEE**.
Don't waste materials.	No gaste materiales.	No **GAS**-tey ma-tey-ree-**ALL**-ess.

Common Phrases for Doing the Work

English	Spanish	Pronunciation
Be careful with the saw.	*Cuidado con la cortadora.*	kwe-DA-doe con la core-ta-**DOE**-ra.
Bring me...	*Tráigame...*	TRY-ga-mey...
...the nails.	...*clavos.*	...**CLA**-bos.
...the hammer.	...*el martillo.*	...el mar-**TEE**-yo.
Build this wall with 2x4s.	*Construya esta pared de dos por cuatro.*	con-**STREW**-ya **ES**-ta pa-**RED** dey dose pore **KWA**-tro.
Build the walls here.	*Construya las paredes aquí.*	con-**STREW**-ya las pa-**RED**-ess ah-**KEY**.
Cut the wood straight.	*Corte la madera derecho.*	**CORE**-tey la ma-**DEY**-roe dey-**REY**-cho.
Cut with the circular saw.	*Corte con la sierra circular*	**CORE**-tey con la **SEE-EH**-rrah seer-**COO**-lar.
We'll finish this...	*Terminamos este...*	tear-mee-**NA**-mos **ES**-tey...
...today.	...*hoy.*	...ohy.
...tomorrow.	...*mañana.*	...ma-**NYA**-na.
Install a header beam here.	*Instale un cabezal aquí.*	een-STA-ley oon ca-bes-**SAL** ah-**KE**.

English	Spanish	Pronunciation
Hold it there...	Sostenga allí...	sos-TEN-ga ah-YEE...
...while I nail it.	...mientras yo clavo.	...MEE-EN-tras yo **CLA**-bo.
...and nail it.	...y clave.	...ee **CLA**-bey.
Level it.	Nivélelo.	nee-**BEY**-ley-low.
Lower it a little.	Bájela un poco.	BA-hey-la oon **POE**-co.
Mark it with a pencil.	Marque con el lápiz.	MAR-kay con el **LA**-peace.
We need...	Necesitamos...	ney-sey-see-**TA**-mos...
...more nails.	...más clavos.	...mass **CLA**-bos.
...more lumber.	...más madera.	...mass ma-**DEY**-ra.
Raise it a little.	Levántela un poco.	le-**BAN**-tey-la oon **POE**-co.
Use plywood for the floor.	Use hojas de madera para el piso.	OO-sey **OH**-has dey ma-**DEY**-ra **PA**-ra el **PEE**-so.

Common Phrases for Finishing Up

English	Spanish	Pronunciation
Clean up the dust.	*Limpie el polvo.*	LEEM-pea-eh el **POLE**-bo.
Clean the tools.	*Limpie las herramientas.*	LEEM-pea-eh las eh-rra-**MEE-EN**-tas.
Please do this again.	*Por favor hágalo otra vez.*	pore fa-**BORE** AH-ga-low **OH**-tra bes.
Fix this.	*Arregle esto.*	ah-**RREG**-ley **ES**-toe.
Pick up the...	*Recoja...*	rey-**KO**-ha...
...tools.	*...las herramientas.*	...las eh-rra-**MEE-EN**-tas.
...trash.	*...la basura.*	...la ba-**SUE**-ra.
...leftover materials.	*...el resto de los materiales.*	el **RESS**-toe dey los ma-tey-ree-**ALL**-ess.
Put the materials in the truck.	*Ponga los materiales en el camión.*	PONE-ga los ma-tey-ree-**ALL**-ess en el ca-me-**OWN**.
...the tools	*...las herramientas*	...las eh-rra-**MEE-EN**-tas.

General Work

Commands and Instructions

English	Spanish	Pronunciation
Adjust it a little.	*Ajústelo un poco.*	ah-WHOSE-tey-low oon **POE**-ko.
Begin...	*Empiece...*	em-**PEE**-eh-sey...
...here.	*...aquí.*	...ah-**KEY**.
...over there.	*...allí.*	...ah-**YEE**.
...tomorrow.	*...mañana.*	...ma-**NYA**-na.
Bring it to me.	*Tráigamelo.*	**TRY**-ga-mey low.
Carry...	*Lleve...*	yea-**BEY**...
...the wood.	*...la madera.*	...la ma-**DEY**-rah.
...the tools.	*...los materiales.*	...los ma-tey-ree-**ALL**-es.
...this.	*...esto.*	...**ES**-toe.
Change...	*Cambie...*	...**KAM**-bee-eh...
...the blade.	*...la cuchilla.*	...la coo-**CHEE**-ya.
...the drill bit.	*...la broca.*	...la **BRO**-ka.

English	Spanish	Pronunciation
Clean...	Limpie...	LEEM-pee-eh...
...the tools.	...las herramientas.	...las eh-rra-**MEE-EN**-tas.
...this room.	...este cuarto.	...**ES**-tey **KWAR**-toe.
...the floor.	...el piso.	...el **PEE**-so.
Close it.	Ciérrelo.	see-**EH**-rrey-lo.
Collect...	Recoja...	rey-**KO**-ha...
...the trash.	...la basura.	...la ba-**SUE**-ra.
...the tools.	...las herramientas.	...las eh-rra-**MEE-EN**-tas.
...the leftover materials.	...el resto de los materiales.	...el **RESS**-toe dey los ma-tey-ree-**ALL**-ess.
Come here.	Venga acá.	BEN-ga ah-**KA**.
Count...	Cuente...	KWHEN-tey...
...the 2x4's.	...las dos por cuatro.	...los dose pore **KWA**-tro.
...the plywood.	...las hojas de madera.	...las **OH**-has dey ma-**DEY**-ra.
...the sheets of drywall.	...los tableros de yeso.	...los ta-**BLEY**-ros dey **YEA**-so.
Cover it with plastic.	Cúbralo con plástico.	KOO-bra-low con **PLAS**-tee-ko.
...tightly.	...bien.	...**BEE**-in.

English	Spanish	Pronunciation
Cut it...	*Córtelo...*	CORE-tey-lo...
...at 45 degrees.	*...a cuarenta y cinco grados.*	...ah kwa-**REN**-ta ee **SEEN**-co **GRA**-dose.
...exactly.	*...exacto.*	...ecks-**ACK**-toe.
...straight.	*...derecho.*	...dey-**REY**-cho.
Do it...	*Hágalo...*	AH-ga-lo...
...later.	*...después.*	...des-**PWHES**.
...now.	*...ahora.*	...ah-**ORE**-ah.
...over.	*...otra vez.*	...**OH**-tra bess.
Don't drop it.	*No lo dejes caer.*	no low **DEY**-heys ka-**AIR**.
Empty it.	*Vacíelo.*	bah-**CEE**-eh-low.
Fill it up.	*Llénelo.*	**YEA**-ney-low.
Finish...	*Termine el trabajo...*	tear-**MEE**-ney el tra-**BA**-hoe...
...today.	*...hoy.*	...ohy.
...tomorrow.	*...mañana.*	...ma-**NYA**-na.
Follow me.	*Sígame.*	**SEE**-ga-mey.

English	Spanish	Pronunciation
Give it to me.	*Démelo.*	**DEY**-mey-low.
Go...	*Vaya...*	BYE-ya...
...over there.	*...allá.*	...ah-**YA**.
...to the store.	*...a la tienda.*	...ah la **TEE-EN**-da.
...to the truck.	*...al camión.*	...all ca-mee-**OWN**.
...with him.	*...con él.*	...con **EL**.
Hammer this.	*Martille esto.*	mar-**TEE**-yea **ES**-toe.
Hang it here.	*Colóquelo aquí.*	co-**LOW**-kay-low ah-**KEY**.
Help him...	*Ayúdele...*	ah-**YOU**-dey-ley...
...lift the beam.	*...a levantar la viga.*	...ah ley-ban-**TAR** la **BEE**-ga.
...unload the truck.	*...a descargar el camión.*	...ah des-car-**GAR** el ca-mee-**OWN**.
Help me...	*Ayúdeme...*	ah-**YOU**-dey-mey...
...lift the beam.	*...a levantar la viga.*	...ah ley-ban-**TAR** la **BEE**-ga.
...unload the truck.	*...a descargar el camión.*	...ah des-car-**GAR** el ca-mee-**OWN**.
Get someone to help you.	*Pida ayuda a alguien.*	**PEE**-da ah-**YOU**-da ah **AL**-gee-en.

English	Spanish	Pronunciation
Hold it there...	*Sosténgalo allí...*	sos-TEN-ga-low ah-YEE...
...and nail it.	...*y clávelo.*	...ee **CLA**-bey-low.
...while I nail it.	...*mientras yo clavo.*	...mee-**EN**-tras yo **CLA**-bow.
Hold this.	*Sostenga este.*	sos-TEN-ga **ES**-tey.
Install it.	*Instálelo.*	een-**STA**-ley-low.
Keep it dry.	*Guárdelo seco.*	GWAR-de-low **SEY**-ko.
...wet.	...*húmedo.*	...**OO**-mey-doe.
...covered.	...*cubierto.*	...coo-**BEE-AIR**-toe.
...open.	...*abierto.*	...ah-**BEE-AIR**-toe.
...closed.	...*cerrado.*	...sey-**RRA**-doe.
Level it.	*Nivélelo.*	nee-**BEY**-ley-low.
Look for...	*Busque...*	BOOS-kay...
...the hammer.	...*el martillo.*	...el mar-**TEE**-yo.
...the nails.	...*los clavos.*	...los **CLA**-bose.
...the tool box.	...*la caja de herramientas.*	...la **CA**-ha dey eh-rra-**MEE-EN**-tas.

English	Spanish	Pronunciation
Loosen it.	Aflójelo.	ah-**FLOW**-hey-low.
Lower it...	Bajeló...	BA-hey-low...
...a little.	...un poco.	...oon **POE**-ko.
...more.	...más.	...mass.
Measure...	Mida...	MEE-da...
...the height.	...el alto.	...el **AL**-toe.
...the length.	...el largo.	...el **LAR**-go.
...the width.	...el ancho.	...el **AN**-cho.
Mix...	Mezcle...	MESS-cley...
...the mortar.	...el mortero.	...el more-**TEY**-roe.
...the compound.	...la masilla.	...la ma-**SEE**-ya.
Move...	Mueva...	MOO-EH-ba...
...the materials.	...los materiales.	...los ma-tey-ree-**ALL**-es.
...the tools.	...las herramientas.	...las eh-rra-**MEE-EN**-tas.
...the truck.	...el camión.	...el ca-me-**OWN**.

English	Spanish	Pronunciation
Move it...	Muévalo...	moo-EH-ba-low...
...a little.	...un poco.	...oon **POE**-ko.
...down.	...abajo.	...ah-**BA**-hoe.
...to the left.	...a la izquierda.	...ah la is-key-**AIR**-da.
...to the right.	...a la derecha.	...ah la dey-**REY**-cha.
...up.	...arriba.	...ah-**RREE**-ba.
Nail it.	Clavelo.	CLA-bey-lo.
Open...	Abra...	AH-bra...
...the can.	...la lata.	...la **LA**-ta.
...the bag.	...la bolsa.	...la **BOWL**-sa.
...the door.	...la puerta.	...la **PWHERE**-ta.
...the truck.	...el camión.	...el ka-mee-**OWN**.
Organize...	Organice...	or-ga-**NEE**-sey...
...the materials.	...los materiales.	...los ma-tey-ree-**ALL**-es.
...the tools.	...las herramientas.	...las eh-rra-**MEE-EN**-tas.
...the lumber.	...la madera.	...la ma-**DEY**-ra.

English	Spanish	Pronunciation
Pick up all of these.	*Recoja todo.*	rey-KO-ha **TOE**-doe.
...the trash.	*...la basura.*	...la ba-**SUE**-ra.
...the nails.	*...los clavos.*	...los **CLA**-bose.
...the leftover materials.	*...el resto de los materiales.*	...el **RESS**-toe dey los ma-tey-ree -**ALL**-es.
...the tools.	*...las herramientas.*	...las eh-rra-**MEE-EN**-tas.
Pull it.	*Jale.*	**HA**-ley.
Push it.	*Empuje.*	em-**POO**-hey.
Put...	*Póngalo...*	**PONE**-ga-lo...
...it in the trash.	*...en la basura.*	...en la ba-**SUE**-ra.
...it over there.	*...allá.*	...ah-**YA**.
Put pressure on it.	*Presiónelo.*	prey-see-**OH**-ney-lo.
Raise it...	*Levántelo...*	le-**BAN**-tey-lo...
...a little.	*...un poco.*	...oon **POE**-ko.
...an inch.	*...una pulgada.*	...**OO**-na pool-**GA**-da.

English	Spanish	Pronunciation
Remove the ladder.	Quite la escalera.	KEY-tey la es-ka-LEY-ra.
...the tape.	...la cinta.	...la SEEN-ta.
...the tarp.	...la lona.	...la LOW-na.
...the scaffold.	...el andamio.	...el an-DA-mee-oh.
Replace the blade.	Cambie la cuchilla.	CAM-bee-eh la coo-CHEE-ya.
...the rotten wood.	...la madera podrida.	...la ma-DEY-ra poe-DREE-da.
...the damaged sheets.	...las hojas dañadas.	...las OH-has da-NYA-das.
Sand...	Lije...	LEE-hey...
...with 150.	...con ciento cincuenta.	...con SEE-en-toe seen-KWEN-ta.
...with 220.	...con dos veinte.	...con dose BEN-tey.
...a little more.	...un poco más.	...oon POE-co MAS.
Shovel this into the wheelbarrow.	Cargue esto en la carretilla.	CAR-gey ES-toe en la ka-rrey-TEE-ya.
Show him.	Muéstrele.	MWES-trey-ley.
Show me.	Muéstreme.	MWES-trey-mey.
Squeeze it out.	Sacúdalo.	sa-COO-day-low.

English	Spanish	Pronunciation
Stack the lumber over there.	Apile la madera allí.	ah-PEE-ley la ma-DEY-ra ah-YEE.
...pipes...	...la tubería...	...la too-bey-REE-ah...
...insulation...	...el aislante...	...el eye-ees-LAN-tey...
...bricks...	...los ladrillos...	...los la-DREE-yo...
Sweep this up.	Barra esto.	BA-rra ES-toe.
...the floors.	...los pisos.	...los PEE-sos.
...the sidewalk.	...la acera.	...la ah-SEY-ra.
...the dust.	...el polvo.	...el POLE-bo.
Take this to the truck.	Lleve esto al camión.	YEA-bey ES-toe all ka-mee-OWN.
Tie it up tightly.	Amárrelo bien.	ah-MA-rrey-low BEE-in.
...with rope.	...con cuerda.	...con KWHERE-da.
...with string.	...con hilo.	...con EE-low.
Tighten it.	Apriételo.	ah-pree-EH-tey-low.
Turn it clockwise.	Gírelo a la derecha.	HEE-rey-low a la dey-REY-cha.
...counter-clockwise.	...a la izquierda.	...a la is-key-AIR-da.

English	Spanish	Pronunciation
Turn it over.	*Déle la vuelta.*	DEY-ley la BWELL-ta.
Turn it off.	*Apague.*	ah-PA-gey.
Turn it on.	*Prenda.*	PREN-da.
Use the hammer like this.	*Use el martillo así.*	OO-sey el mar-**TEE**-yo ah-**SEE**.
...pick...	*...el pico...*	...el **PEE**-ko...
...shovel...	*...la pala...*	...la **PA**-la...
...circular saw...	*...la cortadora circular...*	...la core-ta-**DOE**-ra seer-coo-**LAR**...
Wait...	*Espere...*	es-**PEY**-rey...
...five minutes.	*...cinco minutos.*	...**SEEN**-ko mee-**NEW**-toes.
...two hours.	*...dos horas.*	...dose **OR**-ras.
...until it's dry.	*...hasta que se seque.*	...**AHS**-ta kay sey **SEY**-kay.
...until tomorrow.	*...hasta mañana.*	...**AHS**-ta ma-**NYA**-na.
Wash the brushes.	*Lave las brochas.*	LA-bey las **BRO**-chas.
...the tools.	*...las herramientas.*	...las eh-rra-**MEE-EN**-tas.
Don't waste materials.	*No gaste los materiales.*	no GAS-tey los ma-tey-ree-**ALL**-ess.

Insulation

Tools & Equipment

English	Spanish / Pronunciation
Blowing Machine	*la máquina sopladora* la **MA**-key-na so-pla-**DOE**-ra
Dust Mask	*la mascarilla contra polvo* la mass-ka-**REE**-ya **CON**-tra **POLE**-bo
	la careta la ca-**REY**-ta
Extension Cord	*la extensión eléctrica* la ecks-ten-see-**ON** ee-**LECK**-tree-ka
Goggles	*los lentes de protección* los **LEN**-teys dey pro-teck-see-**ON**
Hammer	*el martillo* el mar-**TEE**-yo
Cap (Hat)	*la gorra* la **GO**-rrah
Ladder	*la escalera* la es-ca-**LEY**-ra

English	Spanish / Pronunciation
Staple Gun	*la pistola engrapadora* la pees-**TOE**-la en-gra-pa-**DOE**-ra
Tape Measure	*la cinta métrica* la **SEEN**-tah **MEY**-tree-ka
	el metro el **MEY**-trow
Utility Knife	*el cuchillo multiuso* el koo-**CHEE**-yo mool-tee-**OO**-so
	la navaja la na-**BA**-ha
Work Gloves	*los guantes de trabajo* los **GWAN**-teys dey tra-**BA**-ho
Work Light	*la lámpara de trabajo* la **LAM**-pa-ra dey tra-**BA**-ho

Materials

English	Spanish / Pronunciation
2 x 4	*dos por cuatro* dose pore **KWA**-tro
Air Rafter Vents	*los deflectores* los dey-fleck-**TOE**-reys
Duct Tape	*la cinta adhesiva* la **SEEN**-ta ad-eh-**SEE**-ba
Fiberglass	*la fibra de vidrio* la **FEE**-bra dey **BEED**-ree-oh
Foam	*la espuma* la es-**POO**-ma
Insulation	*el aislante* el eye-ees-**LAN**-tey *el aislamiento* el eye-ees-la-**MEE-EN**-toe
Insulation, Batt of	*el aislante en rollo* el eye-ees-**LAN**-tey en **ROW**-yo
Insulation, Loose Fill	*el aislante celulosa* el eye-ees-**LAN**-tey sey-lou-**LOW**-sah
Insulation with Backing	*el aislante forrado* el eye-ees-**LAN**-tey foe-**RRA**-doe
Insulation without Backing	*el aislante sin forrar* el eye-ees-**LAN**-tey seen foe-**RRAR**
Plywood	*la hoja de madera* la **OH**-ha dey ma-**DEY**-ra
Spackle	*la masilla* la ma-**SEE**-ya
Staples	*las grapas* las **GRA**-pas *las grampas* las **GRAM**-pas
Vapor Barrier	*la barrera de vapor* la ba-**RREY**-rah dey ba-**PORE**

Common Phrases for Setting Up

English	Spanish	Pronunciation
Begin...	*Empiece...*	em-**PEA-A**-sey...
...here.	*...aquí.*	...ah-**KEE**.
...over there.	*...allí.*	...ah-**YEE**.
...in this room.	*...en este cuarto.*	...en **ES**-tey **QUAR**-toe.
Carry this.	*Lleve esto.*	**YEA**-bey **ES**-toe.
Do it...	*Hágalo...*	**AH**-ga-low...
...carefully.	*...con cuidado.*	...con kwee-**DA**-doe.
...like this.	*...así.*	...ah-**SEE**.
Don't do it like this.	*No lo haga así.*	no low **AH**-ga ah-**SEE**.
What else do you need?	*¿Qué más necesita?*	kay **MASS** ney-sey-**SEE**-ta?
We need 15" insulation.	*Necesitamos aislante de quince pulgadas.*	ney-sey-see-**TA**-mose eye-ees-**LAN**-tey dey **KEEN**-sey pull-**GA**-das.
...23"...	*...veintitrés...*	...been-tee-**TRACE**...

94 Insulation

English	Spanish	Pronunciation
Show me what you need.	*Muéstreme que necesita.*	MWES-trey-mey kay ney-sey-**SEE**-ta.
Unload the truck.	*Descargue el camión.*	des-**CAR**-gey el ka-me-**OWN**.
Use a dust mask.	*Use mascarilla.*	**OO**-sey mass-ka-**REE**-ya.
Don't waste materials.	*No gaste materiales.*	no **GAS**-tey ma-tey-ree-**ALL**-ess.
Wear...	*Lleve...*	YEA-bey...
...a long-sleeve shirt.	*...una camisa de manga larga.*	**OO**-na ca-**MEE**-sa dey **MAN**-ga **LAR**-ga.
...long pants.	*...pantalones largos.*	pan-ta-**LONE**-es **LAR**-goes.

Common Phrases for Doing the Work

English	Spanish	Pronunciation
Be careful...	*Cuidado...*	kwe-**DA**-doe...
...with the insulation.	*...con el aislante.*	...kon el eye-ees-**LAN**-tey.
...walking on the joists.	*...al pisar en las vigas.*	...all **PEE**-sar en las **BEE**-gas.

English	Spanish	Pronunciation
Bring me the staple gun.	Tráigame la pistola engrapadora.	TRY-ga-mey la pees-**TOE**-la en-gra-pa-**DOE**-ra.
...a batt of insulation.	...un rollo de aislante.	...oon **ROW**-yo dey eye-ees-**LAN**-tey.
Cover the pipes with this.	Cubra los tubos con esto.	**COO**-bra los **TOO**-bos con **ES**-toe.
Don't cover the vents.	No cubra las rejillas de ventilación.	no **COO**-bra las rey-**HEE**-yas dey ben-tee-la-see-**OH**.
Cut the insulation.	Corte el aislante.	**CORE**-tey el eye-ees-**LAN**-tey.
Fill in the cracks with foam.	Rellene las ranuras con espuma.	rey-**YE**-ney las ra-**NEW**-ras con es-**POO**-ma.
Refill the blowing machine.	Rellene la máquina sopladora.	rey-**YE**-ney la **MA**-key-na so-pla-**DOE**-ra.
We'll finish this...	Terminamos esto...	tear-mee-**NA**-mos **ES**-toe.
...today.	...hoy.	...ohy.
...tomorrow.	...mañana.	...ma-**NYA**-na.
We need more nails.	Necesitamos más grapas.	ney-sey-see-**TA**-mos mass **GRA**-pas.
...more insulation.	...más aislante.	...mass eye-ees-**LAN**-tey.
Put air rafter vents here.	Ponga deflectores aquí.	**PONE**-ga dey-fleck-**TOE**-reys ah-**KEY**.
Put down two coats.	Ponga dos capas.	**PONE**-ga dose **KA**-pas.

Insulation 96

English	Spanish	Pronunciation
Put the insulation between...	*Ponga aislante entre...*	PONE-ga eye-ees-**LAN**-tey **EN**-trey...
...the joists.	*...las vigas.*	...las **BEE**-gas.
...the studs.	*...los montantes.*	...los mon-**TAN**-teys.
...the 2x4's.	*...los dos por cuatro.*	...los dose pore **KWA**-tro.
Don't put insulation here.	*No ponga el aislante aquí.*	no PONE-ga el eye-ees-**LAN**-tey ah-**KEY**.
Use this insulation...	*Use este aislante...*	OO-sey **ES**-ta eye-ees-**LAN**-tey...
...here.	*...aquí.*	...ah-**KEY**.
...on the walls.	*...en las paredes.*	...en las pa-**RED**-ess.
...in the ceiling.	*...en el techo.*	...en el **TEY**-cho.

Common Phrases for Finishing Up

English	Spanish	Pronunciation
Clean up the dust.	*Limpie el polvo.*	LEEM-pea-eh el **POLE**-bo.
Please do this again.	*Por favor hágalo otra vez.*	pore fa-**BORE** AH-ga-low **OH**-tra bes.

English	Spanish	Pronunciation
When will you finish?	*¿Cuando terminará?*	**KWAN**-doe tear-mee-nah-**RA**?
Pick up the...	*Recoja...*	rey-**KO**-ha...
...tools.	*...las herramientas.*	...las eh-rra-**MEE-EN**-tas.
...trash.	*...la basura.*	...la ba-**SUE**-ra.
...leftover materials.	*...el resto de los materiales.*	...el **RESS**-toe dey los ma-tey-ree-**ALL**-ess.
Put the materials in the truck.	*Ponga los materiales en el camión.*	**PONE**-ga los ma-tey-ree-**ALL**-ess.
Put it in the...	*Póngalo en...*	**PONE**-ga-lo en...
...trash.	*...la basura.*	...la ba-**SUE**-ra.
...truck.	*...el camión.*	...el ca-me-**OWN**.
Sweep the floors.	*Barra los pisos.*	**BA**-rra los **PEE**-sos.
Vacuum...	*Aspire...*	as-**PEE**-rey...
...everything.	*...todo.*	...**TOE**-doe.
...in here.	*...por aquí.*	...pore ah-**KEY**.
...up the dust.	*...el polvo.*	...el **POLE**-bow.
Fix this.	*Arregle esto.*	ah-**RREG**-ley **ES**-toe.

Insulation 98

Landscaping

Tools & Equipment

English	Spanish / Pronunciation
Axe	el hacha
	el **AH**-cha
Blade (mower)	la navaja
	la na-**BA**-ha
Blower	el soplador
	el sow-pla-**DOOR**
Broom	la escoba
	la es-**CO**-ba
Saw, Chain	la sierra de cadena
	la **SEE-EH**-rra de ca-**DEY**-na
Saw, Hand	el serrucho
	el sey-**RREW**-cho
Hoe	el azadón
	el ah-sa-**DON**

English	Spanish / Pronunciation
Hose	la manguera
	la man-**GEY**-ra
Pitch Fork	la horca
	la **OR**-ka
Pruners	la podadora
	la po-da-**DOE**-ra
Spade	la pala
	la **PA**-la
Rake	el rastrillo
	el ras-**TREE**-yo
Rake, Leaf	el rastrillo para hojas
	el ras-**TREE**-yo **PA**-ra **OH**-has
Rake, Yard	el rastrillo para césped
	el ras-**TREE**-yo **PA**-ra **CESS**-ped

English	Spanish / Pronunciation
Roto-tiller	*el aflojador de tierra*
	el ah-flow-ha-**DOOR** dey **TEE-EH**-rra
Shovel	*la pala*
	la **PA**-la
Shovel, Flat	*la pala plana*
	la **PA**-la **PLA**-na
Tractor	*el tractor*
	el track-**TORE**
Trimmers, Hand	*la podadora*
	la po-da-**DOE**-ra

English	Spanish / Pronunciation
Trimmers, Hedge	*la cortadora de seto*
	la core-ta-**DOE**-ra dey **SEY**-toe
Weed Eater	*el desyerbador*
	el des-yer-ba-**DOOR**
	la recortadora giratoria
	la rey-core-ta-**DOE**-ra hee-ra-**TOE**-ree-ah
Wheelbarrow	*la carretilla*
	la ca-rrey-**TEA**-ya

Materials

English	Spanish / Pronunciation
Bag	*la bolsa* la **BOWL**-sa
Bark Chips	*los pedazos de corteza* los pey-**DA**-sos dey core-**TEY**-sa
Bud (of a leaf)	*el brote (de hoja)* el **BRO**-tey
Bud (of a flower)	*el capullo* el ka-**POO**-yo
Bulb	*el bulbo* el **BOOL**-bo
Bush	*el arbusto* el are-**BOOS**-toh

English	Spanish / Pronunciation
Brick	*el ladrillo* el la-**DREE**-yo
	el tabique el ta-**BEE**-kay
Fertilizer	*el abono* el ah-**BOW**-no
	el fertilizante el fer-tee-lee-**SAN**-tey
Flowers	*las flores* las **FLOOR**-ehs
Garden, Vegetable	*la huerta* la **WHERE**-ta
Gloves	*los guantes* los **GWAN**-teys

English	Spanish / Pronunciation
Grass	el césped
	el **CESS**-ped
	el pasto
	el **PAS**-toe
Gravel	la grava
	la **GRA**-ba
Herbicide	el herbicida
	el air-bee-**SEE**-da
Insecticide	el insecticida
	el een-seck-tee-**SEE**-da
Ivy	la hiedra
	la **EE-EH**-dra
Lights	las luces
	las **LOO**-seys
Manure	el estiércol
	el es-tee-**AIR**-cole

English	Spanish / Pronunciation
Plastic	el plástico
	el **PLAS**-tee-ko
Poison	el veneno
	el be-**NEY**-no
Potting Soil	la tierra abonada para jardín
	la **TEE-EH**-rra ah-bow-**NA**-da **PA**-ra har-**DEEN**
Rocks	las piedras
	las **PEE-EH**-dras
Sand	la arena
	la ah-**REY**-na
Seed	la semilla
	la sey-**MEE**-ya
Seed, Grass	la semilla para césped
	la sey-**MEE**-ya **PA**-ra **CESS**-ped

English	Spanish / Pronunciation
Shrub	*el arbusto*
	el are-**BOOS**-toe
Sod	*el césped en rollo*
	el **CESS**-ped en **ROE**-yo
	el tepe
	el **TEY**-pey
Soil	*el suelo*
	el **SWEY**-low
	la tierra
	la **TEE-EH**-rra

English	Spanish / Pronunciation
Straw	*la paja*
	la **PA**-ha
Yard	*la yarda*
	la **YAR**-da
	el patio
	el **PA**-tee-oh

Common Phrases for Setting Up

English	Spanish	Pronunciation
Begin...	Empiece...	em-PEA-A-sey...
...here.	...aquí.	...ah-**KEE**.
...over there.	...allí.	...ah-**YEE**.
...in the back.	...atrás.	...ah-**TRAS**.
...in the front.	...adelante.	...ah-dey-**LAN**-tey.
Carry this.	Lleve esto.	YEA-bey **ES**-toe.
Do it carefully.	Hágalo con cuidado.	AH-ga-low con kwee-**DA**-doe.
...this way.	...así.	...ah-**SEE**.
...slower.	...más despacio.	...mass des-**PA**-see-oh.
Don't do it like this.	No lo haga así.	no low **AH**-ga ah-**SEE**.
What else do you need?	¿Qué más necesita?	kay mass ne-sey-SEE-ta?
Show me what you need.	Muéstreme que necesita.	MWES-trey-mey kay ne-sey-**SEE**-ta.
Unload the truck.	Descargue el camión.	des-**CAR**-gey el ka-me-**OWN**.

Common Phrases for Doing the Work

English	Spanish	Pronunciation
Apply insecticide...	*Aplique insecticida...*	ah-**PLEE**-key een-seck-tee-**SEE**-da...
...to the flower.	*...a las flores.*	...ah las **FLOOR**-ehs.
...to the plants.	*...a las plantas.*	...ah las **PLAN**-tas.
...to the trees.	*...a los árboles.*	...ah los **ARE**-bowl-es.
Bring me...	*Tráigame...*	**TRY**-ga-mey...
...a shovel.	*...la pala.*	...la **PA**-la.
...the fertilizer.	*...el abono.*	...el ah-**BOW**-no.
Build a retaining wall here.	*Construya un muro de contención aquí.*	con-**STREW**-ya oon **MOO**-row dey con-ten-see-**ON** ah-**KEY**.
Cover this area with	*Cubra esta área con...*	**COO**-bra **ES**-ta **AH**-rey-ah con...
...plastic.	*...plástico.*	...**PLAS**-tee-ko.
...straw.	*...paja.*	...**PA**-ha.

English	Spanish	Pronunciation
Cut/Mow the lawn...	Corte el césped...	CORE-tey el **CESS**-ped...
...in the front.	...adelante.	...ah-dey-**LAN**-tey.
...in the back.	...atrás.	...ah-**TRAS**.
...on the sides.	...a los lados.	...ah los **LA**-dose.
Dig a hole...	Cave un hoyo...	CA-bey oon **OH**-yo...
...three feet deep.	...a tres pies de profundidad.	...a trace **PEE**-ace dey pro-foon-dee-**DAD**.
...to plant the tree.	...para plantar el árbol.	**PA**-ra plan-**TAR** el **ARE**-bowl.
Dig a trench...	Cave una zanja...	CA-bey **OO**-na **SAN**-ha...
...here.	...aquí.	...ah-**KEY**.
...over there.	...allí.	...ah-**YEE**.
This needs better drainage.	Necesita mejor drenaje.	ney-sey-**SEE**-ta **MEY**-hore drey-**NA**-hey.
Edge...	Orille...	oh-**REE**-yeh...
...the lawn.	...el césped.	...el **CESS**-ped.
...the driveway.	...la entrada.	...la en-**TRA**-da.
...the sidewalk.	...las aceras.	...las ah-**SEY**-ras.

English	Spanish	Pronunciation
Fertilize...	Abone...	ah-**BOW**-ney...
...the lawn.	...el césped.	...el **CESS**-ped.
...the flowers.	...las flores.	...las **FLOOR**-ehs.
...the plants.	...las plantas.	...las **PLAN**-tas.
Lay sod in the front.	Ponga tepe adelante.	PONE-ga **TEY**-pey ah-dey-**LAN**-tey.
...the back.	...atrás.	...ah-**TRAS**.
Mix with mulch.	Mezcle con estiércol.	MESS-cley con es-tee-**AIR**-cole.
I need...	Necesito...	ney-sey-**SEE**-toe...
...the trimmers.	...el podador.	...el poe-da-**DOOR**.
...the hose.	...la manguera.	...la man-**GEY**-ra.
Do you need more...?	¿Necesita más...?	ney-sey-**SEE**-tah mass...?
...fertilizer?	...abono?	...ah-**BOW**-no?
...sand?	...arena?	...ah-**REY**-na?
...help?	...ayuda?	...ah-**YOU**-da?

English	Spanish	Pronunciation
Plant the flowers.	*Plante las flores.*	PLAN-tey las **FLOOR**-ehs.
...the plants.	*...las plantas.*	...las **PLAN**-tas.
...the trees.	*...los árboles.*	...los **ARE**-bowl-es.
Plant the flowers...	*Plante las flores...*	PLAN-tey las **FLOOR**-ehs...
...here.	*...aquí.*	...ah-**KEY**.
...in a line.	*...en línea.*	...en **LEE**-ney-ah.
...three inches apart.	*...cada tres pulgadas.*	**CA**-da trace pull-**GA**-das.
Prune/Trim...	*Pode...*	PO-dey...
...the bush(es).	*...el arbusto / los arbustos.*	...el are-**BOOS**-toe / los are-**BOOS**-tose.
...the tree(s).	*...el árbol / los árboles.*	...el **ARE**-bowl / los **ARE**-bowl-es.
...the branches.	*...las ramas.*	...las **RA**-mas.
Pull the weeds.	*Desyerbe.*	des-**YER**-bey.
Put mulch...	*Ponga estiércol...*	PONE-ga es-tee-**AIR**-cole...
...around the trees.	*...alrededor de los árboles.*	al-rey-**DEY**-door de los **ARE**-bowl-es.
...in the planters.	*...en las macetas.*	...en las ma-**SEY**-tas.
...around the flowers.	*...alrededor de las flores.*	...al-rey-**DEY**-door de las **FLOOR**-ehs.

English	Spanish	Pronunciation
Put down two inches of mulch.	Ponga dos pulgadas de estiércol.	PONE-ga dose pull-GA-das es-tee-AIR-cole.
...three...	...tres...	...trace...
...four...	...cuatro...	...KWA-trow...
Put down four inches of pine straw.	Ponga cuatro pulgadas de paja de pino.	PONE-ga KWA-trow pull-GA-das dey PA-ha dey PEE-no.
...two...	...dos...	...dose...
...three...	...tres...	...trace...
Put the plants here.	Ponga las plantas aquí.	PONE-ga las PLAN-tas ah-KEY.
...fertilizer...	...el abono...	...el ah-BOW-no...
...gravel...	...la grava...	...la GRA-ba...
...sod...	...el césped en rollo...	...el CESS-ped en ROE-yo...
...a tree...	...el árbol...	...el ARE-bowl...
Rake the leaves.	Rastrille las hojas secas.	ras-TREE-yea las OH-has SEY-kas.
...the yard.	...el césped.	...el CESS-ped.

English	Spanish	Pronunciation
Remove the flowers.	*Quite las flores.*	KEY-tey las **FLOOR**-ehs.
...the dead leaves.	*...las hojas muertas.*	...las **OH**-has **MWHERE**-tas.
Remove the rocks from the soil.	*Quite las piedras de la tierra.*	KEY-tey las **PEE-EH**-dras dey la **TEE-EH**-rra.
Repair the sprinkler heads.	*Repare los regaderos.*	rey-**PA**-rey los rey-ga-**DEY**-rows.
Spray the plants with this.	*Fumigue las plantas con esto.*	foo-**MEE**-gey las **PLAN**-tas con **ES**-toe.
...the flowers...	*...las flores...*	...las **FLOOR**-ehs...
...the lawn...	*...el césped...*	...el **CESS**-ped...
Don't spray the vegetable garden.	*No fumigue la huerta.*	no foo-**MEE**-gey la **WHERE**-ta.
Till the soil.	*Cave el suelo.*	CA-bey el **SWEY**-low.
Turn off...	*Apague...*	an-**PA**-gey...
...the lights.	*...las luces.*	...las **LOOSE**-ehs.
...the sprinklers.	*...los regaderos.*	...los rey-ga-**DEY**-rows.
Turn off the water.	*Cierre la llave del agua.*	SEE-air-rey la **YA**-bey dell **AH**-gwa.

English	Spanish	Pronunciation
Turn on...	Prenda...	PREN-dah...
...the lights.	...las luces.	...las **LOOSE**-ehs.
...the sprinklers.	...los regaderos.	...los rey-ga-**DEY**-rows.
Turn on the water.	Abra la llave del agua.	**AH**-bra la **YA**-bey dell **AH**-gwa.
Water...	Riegue agua...	ree-**EH**-gey **AH**-gwa...
...the plants.	...a las plantas.	...ah las **PLAN**-tas.
...the lawn.	...al césped.	...al **CESS**-ped.
...the trees.	...a los árboles.	...ah los **ARE**-bowl-ess.
...the flowers.	...a las flores.	...ah las **FLOOR**-ehs.
Water every other day.	Riegue agua cada dos días.	ree-**EH**-gey **AH**-gwa **KA**-da dose **DEE**-ahs.
Don't over water.	No riegue demasiado.	no ree-**EH**-gey de-mas-**EE**-ah-doe.

Common Phrases for Finishing Up

English	Spanish	Pronunciation
Clean up the trash.	*Limpie la basura.*	LEEM-pea-eh la ba-**SUE**-ra.
Clean the tools.	*Limpie las herramientas.*	LEEM-pea-eh las eh-rra-**MEE-EN**-tas.
Please do this again.	*Por favor hágalo otra vez.*	pore fa-**BORE** AH-ga-low **OH**-tra bes.
When will you finish?	*¿Cuando terminará?*	**KWAN**-doe tear-mee-nah-**RA**?
We'll finish this...	*Terminamos esto...*	tear-mee-**NA**-mos **ES**-toe...
...today.	...*hoy.*	...ohy.
...tomorrow.	...*mañana.*	...ma-**NYA**-na.
Pick up the...	*Recoja...*	rey-**KO**-ha...
...tools.	...*las herramientas.*	...las eh-rra-**MEE-EN**-tas.
...trash.	...*la basura.*	...la ba-**SUE**-ra.
...leftover materials.	...*el resto de los materiales.*	...el **RESS**-toe dey los ma-tey-ree-**ALL**-ess.
Put the materials in the truck.	*Ponga los materiales en el camión.*	PONE-ga los ma-tey-ree-**ALL**-ess en el ca-me-**OWN**.
...the tools...	...*las herramientas...*	...las eh-rra-**MEE-EN**-tas.

English	Spanish	Pronunciation
Put it in the...	*Póngalo en...*	PONE-ga-lo en...
...trash.	*...la basura.*	...la ba-**SUE**-ra.
...truck.	*...el camión.*	...el ca-me-**OWN**.
...dump truck.	*...el camión de volteo.*	...el ca-me-**OWN** dey bowl-**TEY**-oh.
Sweep...	*Barra...*	BA-rra...
...the driveway.	*...la entrada.*	...la en-**TRA**-da.
...the deck.	*...la terraza.*	...la tey-**RRA**-sa.
...the sidewalk.	*...la acera.*	...la ah-**SEY**-ra.
Wash the driveway.	*Lave la entrada.*	LA-bey la en-**TRA**-da.

Masonry

Tools & Equipment

English	Spanish / Pronunciation
Batter Boards	*las marcas* las **MAR**-kas
Bucket	*la cubeta* la coo-**BEY**-ta
Brush	*el cepillo* el cey-**PEE**-yo
Carpenter's Square	*la escuadra* la es-**KWA**-dra
Chalk Line	*el marcador de líneas* el mar-ka-**DOOR** dey **LEE**-ney-ahs *la tira líneas* la **TEE**-ra **LEE**-ney-ahs
Cover	*la tapa* la **TA**-pa
Gloves, Work	*los guantes de trabajo* los **GWAN**-teys dey tra-**BA**-hoe

English	Spanish / Pronunciation
Float	*el flotador* el flow-ta-**DOOR**
Hammer	*el martillo* el mar-**TEE**-yo
Jointer	*el cepillo automático* el sey-**PEE**-yo ow-toe-**MA**-tee-co
Level	*el nivel* el nee-**BELL**
Mixer	*la mezcladora* la mess-cla-**DOE**-ra
Rake	*el rastrillo* el ras-**TREE**-yo
Scaffold	*el andamio* el an-**DA**-mee-oh
Scraper	*el raspador* el ras-pa-**DOOR**

English	Spanish	Pronunciation
Shovel	la pala	la **PA**-la
String Line	el hilo	el **EE**-low
Tape measure	la cinta métrica	la **SEEN**-tah **MEY**-tree-ka
	el metro	a el **MEY**-trow

English	Spanish	Pronunciation
Anchor Bolts	los pernos de anclaje	los **PEAR**-nos dey an-**CLA**-hey
Backfill	la grava	la **GRA**-ba
Bag	la bolsa	la **BOWL**-sa

English	Spanish	Pronunciation
Tile Cutters	la cortadora de loza	la core-ta-**DOE**-ra dey **LOW**-sa
Trowel	la paleta	la pa-**LEY**-ta
	la cuchara de albañil	la coo-**CHA**-ra dey al-ba-**NYEEL**
Wheelbarrow	la carretilla	la ca-rrey-**TEE**-ya

Materials

English	Spanish	Pronunciation
Brick	el ladrillo	el la-**DREE**-yo
	el tabique	el ta-**BEE**-kay
Concrete	el concreto	el con-**KREY**-toe

English	Spanish / Pronunciation
Concrete Block	*el bloque de concreto* el **BLOW**-kay dey con-**KREY**-toe
Concrete Slab	*la losa de concreto* la **LOW**-sa dey con-**KREY**-toe
Footing	*el base de concreto* el **BA**-sey dey con-**KREY**-toe
Foundation	*la fundación* la foon-da-see-**ON**
Foundation Straps	*los sujetadores de metal* los sue-hey-ta-**DOOR**-ehs dey mey-**TAL**
Grout	*el mortero para juntas* el more-**TEY**-row **PA**-ra **WHOON**-tas *la lechareada* la ley-cha-rey-**AH**-da

English	Spanish / Pronunciation
Lime	*la cal* la kal
Mortar	*el mortero* el more-**TEY**-ro
Post	*el poste* el **POSE**-tey
Rebar	*la barilla* la ba-**REE**-ya
Sand	*la arena* la ah-**REY**-na
Stone	*la piedra* la **PEE-EH**-dra
Water	*el agua* el ah-**GWA**

Common Phrases for Setting Up

English	Spanish	Pronunciation
Add...	Agregue...	ah-GREY-gey...
...another bag.	...otra bolsa.	...OH-tra BOWL-sa.
...half a bag.	...media bolsa.	...MEY-dee-ah BOWL-sa.
...two bags.	...dos bolsas.	...dose BOWL-sas.
...a whole bag.	...una bolsa entera.	...OO-na BOWL-sa en-TEY-ra.
Add...	Agregue...	ah-GREY-gey...
...less.	...menos.	...MEY-nos.
...more.	...más.	...mass.
...water.	...agua.	...AH-gwa.
Begin...	Empiece...	em-PEA-A-sey...
...here.	...aquí.	...ah-KEE.
...over there.	...allí.	...ah-YEE.

English	Spanish	Pronunciation
We are going to build....	*Vamos a construir....*	**BA**-mose ah con-strew-**EAR**....
...a column.	*...una columna.*	...**OO**-na co-**LOOM**-na.
...a fireplace.	*...una chimenea.*	...**OO**-na chee-**MEY**-ney-ah.
...a hearth.	*...un hogar de chimenea.*	...oon **OH**-gar dey chee-**MEY**-ney-ah.
...a mailbox.	*...un buzón.*	...oon boo-**SAWN**.
...a patio.	*...un patio.*	...oon **PA**-tee-oh.
...a retaining wall.	*...un muro de contención.*	...oon **MOO**-row dey con-ten-see-**ON**.
...a wall.	*...una pared.*	...**OO**-na pa-**RED**.
...steps.	*...los escalones.*	...los es-ca-**LOW**-nes.
Carry this.	*Lleve esto.*	**YEA**-bey **ES**-toe.
Do it....	*Hágalo....*	**AH**-gah-lo....
...carefully.	*...con cuidado.*	...kon kwee-**DA**-doe.
...slower.	*...más despacio.*	...mass des-**PA**-see-oh.
...this way.	*...así.*	...ah-**SEE**.
Don't do it this way.	*No lo haga así.*	no low **AH**-ga ah-**SEE**.

English	Spanish	Pronunciation
Fill this with...	*Llene con...*	YEA-ney con...
...concrete.	...*concreto*.	...con-**KREY**-tow.
...mortar.	...*mortero*.	...more-**TEY**-row.
...sand.	...*arena*.	...ah-**REY**-na.
You need...	*Necesita...*	ney-sey-**SEE**-ta...
...foundation.	...*la fundación*.	...la foon-da-see-**ON**.
...rebar.	...*barilla*.	...ba-**REE**-ya.
...foundation straps.	...*sujetadores de metal*.	...sue-hey-ta-**DORE**-ehs dey mey-**TAL**.
Show me what you need.	*Muéstreme que necesita.*	MWES-trey-mey kay ne-sey-**SEE**-ta.
Stack the bricks...	*Apile los ladrillos...*	ah-**PEE**-ley los la-**DREE**-yos.
...here.	...*aquí*.	...ah-**KEY**.
...over there.	...*allí*.	...ah-**YEE**.
Don't waste materials.	*No gaste materiales.*	No **GAS**-tey ma-tey-ree-**ALL**-ess.

Common Phrases for Doing the Work

English	Spanish	Pronunciation
Be careful with the bricks.	*Cuidado con los ladrillos.*	kwe-DA-doe kon los la-**DREE**-yos.
...the blocks.	*...los bloques.*	...los **BLOW**-kays.
Border it with...	*Bordee con...*	BORE-dey-eh con...
...brick.	*...ladrillo.*	...la-**DREE**-yo.
...concrete.	*...concreto.*	...con-**KRE**-toe.
...stone.	*...piedra.*	...PEE-**EH**-dra.
Bring me...	*Tráigame...*	TRY-ga-mey...
...the brush.	*...el cepillo.*	...el sey-**PEE**-yo.
...the mortar.	*...el mortero.*	...el more-**TEY**-roe.
...the sand.	*...la arena.*	...la ah-**REY**-na.
Build up five courses.	*Ponga cinco niveles.*	PONE-ga **SEEN**-co nee-**BEY**-les.
...ten...	*...diez...*	...**DEE**-ace.
...twelve...	*...doce...*	...**DOE**-sey.

English	Spanish	Pronunciation
Check the plumb line.	*Chequée la plomada.*	che-kay-**EH** la plough-**MA**-da.
...the sides.	*...los lados.*	...los **LA**-dose.
Cut the brick...	*Corte el ladrillo...*	**CORE**-tey el la-**DREE**-yo...
...at an angle.	*...al ángulo.*	...al **AN**-goo-low.
...straight.	*...derecho.*	...dey-**REY**-cho.
Don't get the brick face dirty.	*No ensucie el frente del ladrillo.*	no en-**SUE**-see-eh el **FREN**-tey dell la-**DREE**-yo.
Hold it...	*Sostenga...*	sos-**TEN**-ga...
...here.	*...aquí.*	...ah-**KEY**.
...steady.	*...fijamente.*	...fee-ha-**MEN**-tey.
Level it.	*Nivélelo.*	nee-**BEY**-ley-lo.
Move it **back**.	*Muévalo atrás.*	**MOO**-eh-ba ah-**TRAS**.
...**forward**.	*...al frente.*	...al **FREN**-tey.
...**up**.	*...arriba.*	...ah-**RREE**-ba.
...**down**.	*...abajo.*	...ah-**BA**-hoe.
Move to the other side.	*Mueva al otro lado.*	**MOO**-eh-ba al **OH**-tro **LA**-doe.

English	Spanish	Pronunciation
Do you need more...	¿Necesita más...	ney-sey-SEE-tah mass...
...mortar?	...mortero?	...more-TEY-roe?
...sand?	...arena?	...ah-REY-na?
I need...	Necesito...	ney-sey-SEE-toe...
...the level.	...el nivel.	...el nee-BELL.
...mortar.	...el mortero.	...el more-TEY-row.
...more brick.	...más ladrillo.	...mass la-DREE-yo.
Set it...	Ponga....	PONE-ga...
...flush.	...a ras.	...ah rass.
...in mortar.	...en mortero.	...en more-TEY-roe.
Let it set up.	Deje que se seque.	DEY-hey kay sey SEY-kay.
Snap a chalk line after each course.	Marque una línea después de cada nivel.	MAR-kay OO-na LEE-ney-ah dess-PWES dey CA-da nee-BELL.
Straighten out the bricks.	Arregle los ladrillos.	ah-RREY-gley los la-DREE-yos.
Use a string line.	Use el hilo.	OO-sey el EE-low.

124 *Masonry*

Common Phrases for Finishing Up

English	Spanish	Pronunciation
Clean up the dust.	*Limpie el polvo.*	LEEM-pea-eh el **POLE**-bo.
Clean the tools.	*Limpie las herramientas.*	LEEM-pea-eh las eh-rra-**MEE-EN**-tas.
Please do this again.	*Por favor hágalo otra vez.*	pore fa-**BORE** AH-ga-low **OH**-tra bes.
When will you finish?	*¿Cuando terminará?*	**KWAN**-doe tear-mee-nah-**RA**?
We'll finish this...	*Terminamos esto...*	tear-mee-**NA**-mos **ES**-toe...
...today.	...*hoy.*	...ohy.
...tomorrow.	...*mañana.*	...ma-**NYA**-na.
Fix this.	*Arregle esto.*	ah-**RREG**-ley **ES**-toe.
Pick up the...	*Recoja...*	rey-**KO**-ha...
...tools.	...*las herramientas.*	...las eh-rra-**MEE-EN**-tas.
...trash.	...*la basura.*	...la ba-**SUE**-ra.
...leftover materials.	...*el resto de los materiales.*	...el **RESS**-toe dey los ma-tey-ree-**ALL**-ess.

English	Spanish	Pronunciation
Put the materials in the truck.	Ponga los materiales en el camión.	PONE-ga los ma-tey-ree-**ALL**-ess en el ca-me-**OWN**.
...the tools...	...las herramientas...	...las eh-rra-**MEE-EN**-tas.
Put it in the...	Póngalo en...	PONE-ga-lo en...
...trash.	...la basura.	...la ba-**SUE**-ra.
...truck.	...el camión.	...el ca-me-**OWN**.
Sweep the sidewalk.	Barra la acera.	BA-rra la ah-**SEY**-ra.
Touch-up.	Chequée las fallas.	che-kay-**EH** las **FA**-yas.

Painting

Tools & Equipment

English	Spanish / Pronunciation
Bucket	*la cubeta*
	la coo-**BEY**-ta
Compressor	*el compresor*
	el com-prey-**SORE**
Drop Cloth	*la manta (para proteger el piso)*
	la **MAN**-ta (**PA**-ra pro-**TEY**-hair el **PEE**-so)
Extension Pole	*el telescopio*
	el tey-less-**KO**-pee-oh
Ladder	*la escalera*
	la es-kah-**LAIR**-a
Paint Tray	*la bandeja*
	la ban-**DEY**-ha
Paintbrush	*la brocha*
	la **BRO**-cha
Paintbrush, Angled	*la brocha en ángulo*
	la **BRO**-cha en **AN**-goo-low

English	Spanish / Pronunciation
Paintbrush, 3-inch	*la brocha de tres pulgadas*
	la **BRO**-cha dey trace pull-**GA**-das
Paint Sprayer	*la pistola de pintar*
	la pees-**TOE**-la dey peen-**TAR**
Putty Knife	*la espátula*
	la es-**PA**-too-la
Rags	*los trapos*
	los **TRA**-pose
Roller	*el rodillo*
	el ro-**DEE**-yo
Rubber Gloves	*los guantes de caucho*
	los **GWAN**-teys dey **COW**-cho
Scaffold	*el andamio*
	el an-**DA**-mee-oh

128 Painting

Materials

English	Spanish / Pronunciation
Masking Tape	*la cinta adhesiva*
	la **SEEN**-ta ad-eh-**SEE**-ba
Paint	*la pintura*
	la peen-**TOO**-ra
Paint, Flat	*la pintura lisa*
	la peen-**TOO**-ra **LEE**-sa
Paint, Gloss	*la pintura brillante*
	la peen-**TOO**-ra bree-**YAN**-tey
Paint, High-gloss	*la pintura extra brillante*
	la peen-**TOO**-ra **ECKS**-tra bree-**YAN**-tey
Paint, Latex	*la pintura latex*
	la peen-**TOO**-ra **LA**-tex
Paint, Oil-based	*la pintura al aceite*
	la peen-**TOO**-ra all ah-**SAY**-tey

English	Spanish / Pronunciation
Paint, Satin	*la pintura satinada*
	la peen-**TOO**-ra sa-tee-**NA**-da
Paint, Semi-gloss	*la pintura semi-brillante*
	la peen-**TOO**-ra **SEY**-mee bree-**YAN**-tey
Paint Remover	*el disolvente*
	el dee-sol-**BEN**-tey
Paint Stripper	*el removedor de pintura*
	el rey-moe-bey-**DOOR** dey peen-**TOO**-ra
Primer	*el sellador*
	el sey-ya-**DOOR**
Spackle	*la masilla*
	la ma-**SEE**-ya
Tarp	*la lona*
	la **LOW**-na

English	Spanish / Pronunciation
Trash Bags	*las bolsas para basura* las **BOWL**-sas **PA**-ra ba-**SUE**-ra
Turpentine	*el aguarrás* el ah-gwa-**RRAS**

English	Spanish / Pronunciation
Varnish	*el barniz* el bar-**NEES** *la laca* la **LA**-ka
Window Glazing	*la masilla para vidrios* la ma-**SEE**-ya **PA**-ra **BEED**-ree-ohs

Common Phrases for Setting Up

English	Spanish	Pronunciation
Begin...	Empiece...	em-**PEA**-A-sey...
...here.	...aquí.	...ah-**KEE**.
...in this room.	...en este cuarto.	...en **ES**-tey **QUAR**-toe.
Clean the wall first.	Limpie la pared primero.	**LEEM**-pee-eh la pa-**RED** pree-**MEY**-row.
Cover the door frames with tape.	Cubra los marcos con cinta adhesiva.	**COO**-bra los **MAR**-kos con **SEEN**-ta ad-eh-**SEE**-ba.
...windows and molding...	...las ventanas y molduras...	...las ben-**TA**-nas ee mole-**DO**-ras...
Cover the floor with tarp.	Cubra los pisos con la lona.	**COO**-bra los **PEE**-sos con la **LOW**-na.
Cover the furniture with plastic.	Cubra los muebles con plástico.	**COO**-bra los **MWEH**-bleys con **PLAS**-tee-ko.
Do you have rollers?	¿Tiene rodillos?	**TEE-EN**-ey ro-**DEE**-yos?
...a scaffold?	...andamio?	...an-**DA**-mee-oh?
Paint the molding.	Pinte las molduras.	**PEEN**-tey las mole-**DO**-ras.
Paint with a brush.	Pinte con brocha.	**PEEN**-tey con **BRO**-cha.
...with the sprayer.	...a soplete.	...ah so-**PLEY**-tey.

English	Spanish	Pronunciation
Can you paint with a sprayer?	¿Puede pintar a soplete?	PWEH-dey peen-**TAR** ah so-**PLEY**-tey?
Patch the holes with spackle.	Emparche los huecos con masilla.	PAR-che los **WHEY**-kos con ma-**SEE**-ya.
Remove the outlet covers.	Quite las tapas de los interruptores.	KEY-tey las **TA**-pahs dey los een-tey-rroup-**TORE**-es.
Remove the loose paint.	Quite la pintura suelta.	KEE-tey la peen-**TOO**-ra **SWELL**-tah.
Sand this wall.	Lije esta pared.	LEE-hey **ES**-tah pa-**RED**.
Stir the paint.	Mezcle la pintura.	MESS-cley la peen-**TOO**-ra.
Take down the light fixtures.	Retire las luces del techo.	rey-TEER-rey las **LOOSE**-ehs del **TEY**-cho.
...the wall paper.	...el papel de tapiz.	...el pa-**PELL** dey ta-**PEACE**.
Tape the windows.	Ponga cinta adhesiva en las ventanas.	PON-ga **SEEN**-ta ad-eh-**SEE**-ba en las ben-**TAH**-nas.

Common Phrases for Doing the Work

English	Spanish	Pronunciation
First, cut in...	*Primero recorte...*	pre-**MEY**-row rey-**CORE**-tey...
...here.	*...aquí.*	...ah-**KEY**.
...around the windows.	*...alrededor de las ventanas.*	...all-rey-dey-**DOOR** dey las ben-**TA**-nas.
...around the cabinets.	*...alrededor de los gabinetes.*	...all-rey-dey-**DOOR** dey los ga-bee-**NEY**-teys.
Do you need more...	*¿Necesita más...*	ney-sey-**SEE**-tah mass...
...paint?	*...pintura?*	...peen-**TOO**-ra?
...primer?	*...sellador?*	...sey-ya-**DOOR**?
Paint the ceiling first.	*Pinte el techo primero.*	**PEEN**-tey el toom-**BA**-doe pree-**MEY**-row.
Paint this room.	*Pinte este cuarto.*	**PEEN**-tey **ES**-tey **QUAR**-toe.
...ceiling.	*...este techo.*	...**ES**-tey **TEY**-cho.
...molding.	*...esta moldura.*	...**ES**-ta mole-**DO**-ra.
...door.	*...esta puerta.*	...**ES**-ta **PWHERE**-ta.
Paint the handrails.	*Pinte los pasamanos.*	**PEEN**-tey los pa-sa-**MA**-nos.
Paint two coats.	*Pinte dos capas.*	**PEEN**-tey dos **KA**-pas.

Painting 133

English	Spanish	Pronunciation
Use a thick coat of paint.	*Use una capa gruesa de pintura.*	OO-sey OO-na KA-pa **GREW-EH**-sa dey pee-TOO-ra.
Re-paint this wall.	*Pinte esta pared otra vez.*	PEEN-tey **ES**-ta pa-**RED OH**-tra beys.
This is wet paint.	*Está recién pintado.*	es-**TA** rey-see-**EN** peen-**TA**-do.
Put paint in the tray.	*Ponga la pintura en la bandeja.*	**PONE**-ga la peen-**TOO**-ra en la ban-**DEY**-ha.
Use this paint.	*Use esta pintura.*	OO-sey **ES**-ta peen-**TOO**-ra.
Prime these walls.	*Pinte con sellador estas paredes.*	PEEN-tey con sey-ya-**DOOR ES**-tas pa-**RED**-ess.

Common Phrases for Finishing Up

English	Spanish	Pronunciation
Clean these...	Limpie...	LEEM-pee-eh...
...windows.	...estas ventanas.	...ES-tas ben-TA-nas.
...doors.	...estas puertas.	...ES-tas PWHERE-tas.
...floors.	...estos pisos.	...ES-tos PEE-sos.
Clean the brushes.	Limpie las brochas.	LEEM-pee-eh las BRO-chas.
Use soap and water.	Use agua y jabón.	OO-sey ah-GWA ee ha-BONE.
Use turpentine.	Use aguarrás.	OO-sey ah-gwa-RRAS.
Please do this again.	Por favor hágalo otra vez.	pore fa-BORE AH-ga-low OH-tra bes.
Paint another coat.	Pinte otra capa.	PEEN-tey OH-tra KA-pa.
Put the lid on the paint can.	Tape la pintura.	TA-pey la peen-TOO-ra.
Remove the tape.	Quite la cinta.	KEY-tey la SEEN-ta.
Soak them in turpentine.	Remójelas en aguarrás.	rey-MOE-hey-las en ah-gwa-RRA.

Roofing

Tools & Equipment

English	Spanish / Pronunciation
Air Hose	*la manguera de aire*
	la man-**GEY**-ra dey **EYE**-rey
Axe	*el hacha*
	el **AH**-cha
Blower	*la sopladora*
	la so-pla-**DOE**-ra
Broom	*la escoba*
	la es-**CO**-ba
Chalk Line	*el marcador de líneas*
	el mar-ka-**DOOR** dey **LEE**-ney-ahs
	el tira líneas
	el **TEE**-ra **LEE**-ney-ahs
Compressor	*el compresor*
	el com-prey-**SORE**

English	Spanish / Pronunciation
Conveyor, Portable	*el transportador portátil*
	el trans-pore-ta-**DOOR** pore-**TA**-teel
Crowbar	*la barra*
	la **BA**-rra
	la pata de cabra
	la **PA**-ta dey **CA**-bra
Hammer	*el martillo*
	el mar-**TEE**-yo
Hammer, Roofing	*el martillo para techar*
	el mar-**TEE**-yo **PA**-ra tey-**CHAR**
Hammer Stapler	*el martillo de grapas*
	el mar-**TEE**-yo dey **GRA**-pas
Hawk Blade	*la navaja de curva*
	la na-**BA**-ha dey **CORE**-ba

English	Spanish / Pronunciation
Ladder	*la escalera*
	la es-ca-**LEY**-ra
Level	*el nivel*
	el nee-**BELL**
Magnet	*el imán*
	el ee-**MAN**
Marking Pencil	*el lápiz de color*
	el **LA**-peace dey co-**LORE**
Metal Snips	*las tijeras para metal*
	las tee-**HEY**-ras **PA**-ra mey-**TAL**
Nail Gun	*la clavadora automática*
	la cla-ba-**DOE**-ra ow-to-**MA**-tee-ca
Putty Knife	*la espátula*
	la es-**PA**-too-la

English	Spanish / Pronunciation
Roofing Gun	*la pistola para brea*
	la pees-**TOE**-la **PA**-ra **BREY**-ah
Shingle Eater	*el removedor de tejas*
	el rey-moe-bey-**DOOR** dey **TEY**-has
Trowel	*la llana*
	la **YA**-na
	la espátula
	la es-**PA**-too-la
Utility Knife	*el cuchillo multiuso*
	el coo-**CHEE**-yo mool-tee-**OO**-so

Materials

English	Spanish / Pronunciation
Aluminum	*el aluminio* el ah-loo-**MEE**-nee-oh
Battens	*los listones* los lees-**TONE**-es
Building Paper	*el papel de construcción* el pa-**PELL** dey con-strook-see-**ON**
Copper	*el cobre* el **CO**-brey
Felt Paper	*el papel de fieltro* el pa-**PELL** dey fee-**EL**-tro
	el papel negro el pa-**PELL NEY**-grow
Flashing	*el tapajuntas* el ta-pa-**WHOON**-tas

English	Spanish / Pronunciation
Gravel	*la grava* la **GRA**-ba
Grout	*la lechareada* la ley-cha-rey-**AH**-da
	el mortero de cemento el more-**TEY**-row dey sey-**MEN**-toe
Insulation	*el aislante* el eyes-**LAN**-tey
Metal	*el metal* el mey-**TAL**
Nails	*los clavos* los **CLA**-bos
Nails, Roofing	*los clavos para techo* los **CLA**-bos **PA**-ra **TEY**-cho

English	Spanish / Pronunciation
Plywood	*la hoja de madera*
	la **OH**-ha dey ma-**DEY**-ra
Roof Rolling	*el rollo*
	el **ROW**-yo
Roofing Cement	*el cemento de techar*
	el sey-**MEN**-toe dey tey-**CHAR**
Sheathing	*la cubierta*
	la coo-**BEE-AIR**-ta
Shingles	*las tejas*
	las **TEY**-has
Shingles, Concrete	*las tejas de concreto*
	las **TEY**-has dey con-**CREY**-toe
Shingles, Hip	*las tejas de equina*
	las **TEY**-has dey es-**KEY**-na
Shingles, Slate	*las tejas de pizarra*
	las **TEY**-has dey pee-**SAR**-rrah

English	Spanish / Pronunciation
Shingles, Wood	*las tejas de madera*
	las **TEY**-has dey ma-**DARE**-ah
Silicon	*el silicón*
	el see-lee-**CON**
Tar	*la brea*
	el **BREY**-ah
Tarp	*la lona*
	la **LOW**-na
Tile, Asphalt	*la teja de asfalto*
	la **TEY**-ha dey as-**FAL**-toe
Tile, Clay	*la teja de arcilla*
	la **TEY**-ha dey are-**SEE**-ya
Tile, Roofing	*la teja*
	la **TEY**-ha

Common Phrases for Setting Up

English	Spanish	Pronunciation
Begin...	*Empiece...*	em-PEA-EH-sey...
...here.	*...aquí.*	...ah-**KEE**.
...over there.	*...allí.*	...ah-**YEE**.
...on this side.	*...en este lado.*	...en **ES**-tey **LA**-doe.
Carry this.	*Lleve esto.*	YEA-bey **ES**-toe.
Cut the felt paper.	*Corte el papel de fieltro.*	CORE-tey el pa-**PELL** dey fee-**EL**-tro.
Lay the felt paper first.	*Ponga el papel de fieltro primero.*	PONE-ga el pa-**PELL** dey fee-**EL**-tro **PREE**-mey-row.
Lay the felt paper horizontally.	*Ponga el papel de fieltro horizontal.*	PONE-ga el pa-**PELL** dey fee-**EL**-tro hoe-ree-**SON**-tal.
Lay shingles.	*Ponga las tejas.*	PONE-ga las **TEY**-has.
Mark off the roof.	*Marque el techo.*	MAR-kay el **TEY**-cho.
Measure and draw chalk lines.	*Mida y trace líneas de tisa.*	MEE-da y **TRA**-cey **LEE**-ney-as dey **TEE**-sa.

English	Spanish	Pronunciation
We need...	Necesitamos...	ney-sey-see-**TA**-mos...
...paper.	...papel.	...pa-**PELL**.
...shingles.	...tejas.	...**TEY**-has.
...nails.	...clavos.	...**CLA**-bos.
Put the materials here.	Ponga los materiales aquí.	**PONE**-ga los ma-tey-ree-**ALL**-ess ah-**KEY**.
Put the ladder against the roof.	Ponga la escalera contra el techo.	**PONE**-ga la es-ca-**LEY**-ra **CON**-tra el **TEY**-cho.
Put the tarp here.	Ponga la lona aquí.	**PONE**-ga la **LO**-na ah-**KEY**.
Remove the old shingles.	Quite las tejas viejas.	**KEY**-tey las **TEY**-has **BEE-EH**-has.
Replace this plywood.	Reemplace esta hoja de madera.	rey-em-**PLA**-sey **ES**-ta **OH**-ha dey ma-**DEY**-ra.
Replace these shingles.	Reemplace estas tejas.	rey-em-**PLA**-sey **ES**-tas **TEY**-has.
Sweep the surface first.	Barra la superficie primero.	**BA**-rra la sue-pear-**FEE**-see-eh pree-**MEY**-roe.
Use...	Use...	**OO**-sey...
...a 2x6.	...un dos por seis.	...oon dose pore seys.
...a 2x8.	...un dos por ocho.	...oon dose pore **OH**-cho.
...these nails.	...estos clavos.	...**ES**-tos **CLA**-bos.

Common Phrases for Doing the Work

English	Spanish	Pronunciation
Be careful on the roof.	*Tenga cuidado en el techo.*	TEN-ga kwe-DA-doe en el TEY-cho.
Be careful, the shingles are hot.	*Tenga cuidado, las tejas están calientes.*	TEN-ga kwe-DA-doe, las TEY-has ES-tan ca-lee-EN-teys.
Bring me...	*Tráigame...*	TRY-ga-mey...
...the hammer.	*...el martillo.*	...el mar-TEE-yo.
...more shingles.	*...más tejas.*	...mass TEY-has.
Cover everything up.	*Cubra todo.*	Coo-bra todo.
Cut the plywood.	*Corte la hoja de madera.*	CORE-tey la OH-ha dey ma-DEY-ra.
Get off the roof!	*¡Bájese del techo!*	BA-hey-sey dell TEY-cho!
Nail the drip edge like this.	*Ponga así los clavos en el borde de goteos.*	PONE-ga ah-SEE los CLA-bos en el BORE-dey dey GO-tey-ohs.
Install the roof felt flat.	*Instale el fieltro del techo plano.*	een-STA-ley el fee-EL-tro dell TEY-cho PLA-no.

English	Spanish	Pronunciation
Install the drip edge.	Instale el borde de goteos.	een-STA-ley el **BORE**-dey dey go-**TEY**-ohs.
Lift...	Levante...	ley-**BAN**-tey...
...the plywood.	...la hoja de madera.	...la **OH**-ha dey ma-**DEY**-ra.
...the flashing.	...el tapajuntas.	...el ta-pa- **WHOON**-tas.
...the shingles.	...las tejas.	...las **TEY**-has.
I need...	Necesito...	ney-sey-**SEE**-toe...
...shingles.	...tejas.	...**TEY**-has.
...some nails.	...clavos.	...**CLA**-bos.
Put flashing here.	Ponga el tapajuntas aquí.	PONE-ga el ta-pa-**WHOON**-tas ah-**KEY**.
...shingles...	...las tejas...	...las **TEY**-has...
Put boots on the pipes.	Ponga botas en los tubos.	PONE-ga **BOW**-tas en los **TOO**-bose.
Put the shingles on top of the house.	Ponga las tejas en el techo.	PONE-ga las **TEY**-has en el **TEY**-cho.
Put the ridge shingles here.	Ponga las tejas para los bordes aquí.	PONE-ga las **TEY**-has pay-ra **BOR**-deys ah-**KEY**.

English	Spanish	Pronunciation
Put the ridge vent first.	*Ponga la tapa de ventilar primero.*	PONE-ga la TA-pa dey ben-tee-LAR pree-MEY-ro.
Put one nail per shingle.	*Ponga un clavo en cada teja.*	PONE-ga oon CLA-bo en CA-da TEY-ha.
...two nails...	...*dos clavos*...	...dose CLA-bose...
...three nails...	...*tres clavos*...	...trace CLA-bose...
Don't put felt paper here.	*No ponga papel de fieltro aquí.*	no PONE-ga pa-PELL dey fee-EL-tro ah-KEY.
Use roofing nails.	*Use clavos para techados.*	OO-sey CLA-bos PA-ra tey-CHA-dose.
Don't walk on the shingles.	*No camine por las tejas.*	no ca-MEE-ney pore las TEY-has.

Common Phrases for Finishing Up

English	Spanish	Pronunciation
Clean out the gutters.	*Limpie las cunetas.*	LEEM-pee-eh las coo-NEY-tas.
Please do this again.	*Por favor hágalo otra vez.*	pore fa-BORE AH-ga-low OH-tra bes.
When will you finish?	*¿Cuando terminará?*	KWAN-doe tear-mee-nah-RA?
Fix this.	*Arregle esto.*	ah-RREG-ley ES-toe.
Pick up the nails with the magnet.	*Recoja los clavos con el imán.*	rey-KO-ha los CLA-bose con el ee-MAN.
Put tar here.	*Ponga brea aquí.*	PONE-ga BREY-ah ah-KEY.
Put the materials in the truck.	*Ponga los materiales en el camión.*	PONE-ga los ma-tey-ree-ALL-ess en el ca-me-OWN.
...the tools...	*...las herramientas...*	...las eh-rra-MEE-EN-tas...
Put it in the...	*Póngalo en...*	PONE-ga-lo en...
...trash.	*...la basura.*	...la ba-SUE-ra.
...truck.	*...el camión.*	...el ca-me-ON.
Tar the visible nails.	*Ponga brea sobre los clavos visibles.*	PONE-ga BREY-ah SO-brey los CLA-bos bee-SEE-bleys.

Siding

Tools & Equipment

English	Spanish / Pronunciation
Chalk Line	*el marcador de líneas*
	el mar-ka-**DOOR** dey **LEE**-ney-ahs
	la tira líneas
	la **TEE**-ra **LEE**-ney-ahs
Compressor	*el compresor*
	el com-prey-**SORE**
Folding Tool	*la herramienta plegadora*
	la eh-rra-**MEE-EN**-ta pley-ga-**DOE**-ra
Hammer	*el martillo*
	el mar-**TEE**-yo
Hammer Tacker	*el martillo engrapador*
	el mar-**TEE**-yo en-gra-pa-**DOOR**
Hand Seam	*la dobladora manual*
	la doe-bla-**DOE**-ra ma-**NEW**-all

English	Spanish / Pronunciation
Ladder	*la escalera*
	la es-ca-**LEY**-ra
Level	*el nivel*
	el nee-**BELL**
Nail Gun	*la pistola de clavos*
	la pees-**TOE**-la de **CLA**-bose
	la clavadora automática
	la cla-ba-**DOE**-rah ow-toe-**MA**-tee-ca
Nail Hole Punch	*el punzón para agujero de clavo*
	el poon-**SON PA**-ra ah-goo-**HEY**-row
	dey **CLA**-bow
Saw, Circular	*la sierra circular*
	la **SEE-EH**-rrah seer-coo-**LAR**
Saw, Mitre	*la sierra de retroceso*
	la **SEE-EH**-rrah dey re-tro-**SEY**-so

English	Spanish / Pronunciation
Scaffold	*el andamio*
	el an-**DA**-mee-oh
Snips	*las tijeras para metal*
	las tee-**HEY**-ras **PA**-ra **MEY**-tal
Speed Square	*la escuadra falsa*
	la es-**KWA**-dra **FALL**-sa
Square	*la escuadra*
	la es-**KWA**-dra

English	Spanish / Pronunciation
Tape Measure	*la cinta métrica*
	la **SEEN**-tah **MEY**-tree-ka
	el metro
	el **MEY**-trow
T-Square	*la regla T*
	la **REY**-gla tey
Work Gloves	*los guantes de trabajo*
	los **GWAN**-teys dey tra-**BA**-hoe

Materials

English	Spanish / Pronunciation
Backer Board	*la planta base*
	la **PLAN**-ta **BA**-sey
Caulking	*la masilla*
	la ma-**SEE**-ya
Cedar shingles	*las tejas de cedro*
	las **TEY**-has dey **SEY**-drow
Corner boards	*las esquinas de acabado*
	las es-**KEY**-nas dey ah-ka-**BA**-doe
	las tablas de esquina
	las **TA**-blas dey es-**KEY**-na
Flashing	*el tapajuntas*
	el ta-pa-**WHOON**-tas
Nails	*los clavos*
	los **CLA**-bose

English	Spanish / Pronunciation
Nails, Finish	*los clavos de terminar*
	los **CLA**-bose dey tear-mee-**NAR**
	los clavos sin cabeza
	los **CLA**-bose seen ca-**BEY**-sa
Siding	*el revestimiento*
	el rey-bes-tee-**MEE-YEN**-toe
Siding, Aluminum	*el revestimiento de aluminio*
	el rey-bes-tee-**MEE-YEN**-toe dey ah-loo-**MEE**-nee-oh
Siding, Asbestos	*el revestimiento de asbestos*
	el rey-bes-tee-**MEE-YEN**-toe dey as-**BES**-tose
Siding, Cement Composite	*el revestimiento de fibra de cemento*
	el rey-bes-tee-**MEE-YEN**-toe dey **FEE**-bra dey sey-**MEN**-toe

English	Spanish / Pronunciation
Siding, Vinyl	*el revestimiento de vinil* el rey-bes-tee-**MEE-YEN**-toe dey bee-**NEEL**
Siding, Wood	*el revestimiento de madera* el rey-bes-tee-**MEE-YEN**-toe dey ma-**DEY**-ra
Starter Strip	*el listón* el lees-**TAWN**
Underlayment	*la plancha base* la **PLAN**-cha **BA**-sey

English	Spanish / Pronunciation
Vent, Exhaust	*el orificio de escape* el oh-ree-**FEE**-see-oh dey es-**KA**-pey
Vent, Gable	*la ventilación del alero* la ben-tee-la-see-**ON** dell ah-**LEY**-row
Weather Barrier	*la barrera de agua y de viento* la ba-**RREY**-ra dey **AH**-gwa ee de bee-**EN**-toe

Common Phrases for Setting Up

English	Spanish	Pronunciation
Begin...	Empiece...	em-PEA-A-sey...
...here.	...aquí.	...ah-KEE.
...over there.	...allí.	...ah-YEE.
Carry this.	Lleve esto.	YEA-bey ES-toe.
Do it like this.	Hágalo así.	AH-ga-low ah-SEE.
Don't do it like this.	No lo haga así.	no low AH-ga ah-SEE.
Look at the plans first.	Vea los planes primero.	BEY-ah los PLA-neys pre-MEY-roe.
What else do you need?	¿Qué más necesita?	kay mass ne-sey-SEE-ta?
Show me what you need.	Muéstreme que necesita.	MWES-trey-mey kay ne-sey-SEE-ta.
Unload the truck.	Descargue el camión.	des-CAR-gey el ka-me-OWN.
Don't waste materials.	No gaste materiales.	no GAS-tey ma-tey-ree-ALL-ess.

Common Phrases for Doing the Work

English	Spanish	Pronunciation
Alternate joints on each side.	Alterne juntas en cada lado.	all-TEAR-ney **WHOON**-tas en **CA**-da **LA**-doe.
Bring me...	Tráigame...	TRY-ga-mey...
...a long piece.	...una pieza larga.	...**OO**-na **PEE-EH**-sa **LAR**-ga.
...a short piece.	...una pieza corta.	...**OO**-na **PEE-EH**-sa **CORE**-ta.
...another piece.	...otra pieza.	...**OH**-tra **PEE-EH**-sa.
...the hammer.	...el martillo.	...el mar-**TEE**-yo.
...the nails.	...clavos.	...**CLA**-bose.
Cut it...	Corte...	CORE-tey...
...straight.	...derecho.	...dey-**REY**-cho.
...at an angle.	...a un ángulo.	...ah oon **AN**-goo-low.
Cut this at 45 degrees.	Corte en cuarenta y cinco grados.	CORE-tey en kwa-**REN**-ta ee **SEEN**-co **GRA**-dose.
Cut while he nails.	Corte mientras él clava.	CORE-tey **MEE-EN**-tras **EL CLA**-ba.
Nail while he cuts.	Clave mientras él corta.	**CLA**-ba **MEE-EN**-tras **EL** CORE-ta.

English	Spanish	Pronunciation
We'll finish...	*Terminamos...*	tear-mee-**NA**-mos...
...today.	...*hoy.*	...ohy.
...tomorrow.	...*mañana.*	...ma-**NYA**-na.
Do you need help?	*¿Necesita ayuda?*	ney-sey-**SEE**-ta ah-**YOU**-da?
Get someone to help you.	*Pida a alguien que le ayude.*	**PEE**-da ah al-**GEE**-en kay ley ah-**YOU**-dey.
Help him...	*Ayúdele...*	ah-**YOU**-dey-ley...
...lift the pieces.	...*a levantar las piezas.*	...ah ley-ban-**TAR** las **PEE-EH**-sas.
...move the scaffold.	...*a mover el andamio.*	...ah moe-**BEAR** el an-**DA**-mee-oh.
Help me...	*Ayúdeme...*	ah-**YOU**-dey-mey...
...lift the pieces.	...*a levantar las piezas.*	...ah ley-ban-**TAR** las **PEE-EH**-sas.
...move the scaffold.	...*a mover el andamio.*	...ah moe-**BEAR** el an-**DA**-mee-oh.
Hold it there...	*Sostenga allí...*	sos-**TEN**-ga ah-**YEE**...
...while I nail it.	...*mientras yo clavo.*	...**MEE-EN**-tras yo **CLA**-bo.
...and nail it.	...*y clave.*	...ee **CLA**-bey.

156 Siding

English	Spanish	Pronunciation
Install the siding. ...starter strip.	Instale el revestimiento. ...el listón.	een-STA-ley el rey-bes-tee-MEE-YEN-toe. ...el lease-TAWN.
Install the corner boards first.	Instale las esquinas primero.	een-STA-ley las es-KEY-nas pree-MEY-row.
Leave a one-inch overlap. ...half-inch...	Deje una pulgada pasado de la base. ...media pulgada...	DEY-hey OO-na pool-GA-da pa-SA-doe de la BA-sey. ...MEY-dee-ah pool-GA-da...
Check the level every 10-12 courses.	Cheqeé el nivel cada 10 a 12 hileras.	che-kay-EH el nee-BELL CA-da DEE-ace a DOSE-eh ee-LEY-ras.
Level it.	Nivélelo.	ni-BEY-ley-low.
Make sure it's level.	Cheqeé el nivel.	che-kay-EH el nee-BELL.
Lower it... ...a little. ...an inch.	Bájela... ...un poco. ...una pulgada.	BA-hey-la... ...oon POE-co. ...OO-na pull-GA-da.
Mark it with a pencil.	Marque con el lápiz.	MAR-kay con el LA-peace.

English	Spanish	Pronunciation
Mark each row on the corner board.	Marque en la esquina donde cada hilera se unirá.	MAR-kay en la es-KEY-na DON-dey CA-da ee-LEY-ra sey oo-nee-RA.
Measure twice, cut once.	Mida dos veces, corte una sola vez.	MEE-da dose BEY-seys, CORE-tey OO-na SO-la bes.
Measure and mark the courses.	Mida y trace las tablas.	MEE-da ee TRA-sey las TA-blas.
We need...	Necesitamos...	ney-sey-see-TA-mos...
...more nails.	...más clavos.	...mass CLA-bos.
...more molding.	...más revestimiento.	...mass rey-bes-tee-MEE-YEN-toe.
Nail a half inch from the edge.	Ponga el clavo a una pulgada del filo.	**PONE**-ga el **CLA**-bow ah **OO**-na pool-**GA**-da dell **FEE**-low.
Place nail in the center of the slot.	Ponga el clavo en el centro de la ranura.	PONE-ga el CLA-bow en el SIN-tro dey la ra-NEW-ra.
Nail every six inches.	Ponga clavo cada seis pulgadas.	PONE-ga CLA-bow CA-da seys pull-GA-das.
Raise it...	Levántela...	le-BAN-tey-la...
...a little.	...un poco.	...oon POE-co.
Remove the old siding.	Quite el revestimiento viejo.	KEY-tey el rey-bes-tee-MEE-YEN-toe BEE-EH-hoe.
...damaged...	...dañado.	...da-NYA-doe.

158 *Siding*

English	Spanish	Pronunciation
Save the scrap pieces.	*Guarde los pedazos sobrantes.*	GWAR-dey los pey-DA-sos so-BRAN-teys.
Snap a chalk line every 12 rows.	*Marque una línea cada doce hileras.*	MAR-kay OO-na LEE-ney-ah CA-da DOSE-sey ee-LEY-ras.
Use work gloves.	*Use los guantes de trabajo.*	OO-sey los GWAN-teys dey tra-BA-hoe.
Work from the bottom up.	*Trabaje desde abajo hacia arriba.*	tra-BA-hey DES-dey ah-BA-hoe AH-cee-ah ah-RREE-ba.

Common Phrases for Finishing Up

English	Spanish	Pronunciation
Please do this again.	Por favor hágalo otra vez.	pore fa-**BORE** AH-ga-low **OH**-tra bes.
When will you finish?	¿Cuando terminará?	**KWAN**-doe tear-mee-nah-**RA**?
Fix this.	Arregle esto.	ah-**RREG**-ley **ES**-toe.
Pick up the...	Recoja...	rey-**KO**-ha...
...tools.	...las herramientas.	...las eh-rra-**MEE-EN**-tas.
...trash.	...la basura.	...la ba-**SUE**-ra.
...leftover materials.	...el resto de los materiales.	...el **RESS**-toe dey los ma-tey-ree-**ALL**-ess.
Put the materials in the truck.	Ponga los materiales en el camión.	**PONE**-ga los ma-tey-ree-**ALL**-ess en el ca-me-**OWN**.
...the tools...	...las herramientas...	...las eh-rra-**MEE-EN**-tas.
Put it in the...	Póngalo en...	**PONE**-ga-lo en...
...trash.	...la basura.	...la ba-**SUE**-ra.
...truck.	...el camión.	...el ca-me-**OWN**.